Combined Scheduling and Control

Special Issue Editors

John D. Hedengren
Logan Beal

MDPI • Basel • Beijing • Wuhan • Barcelona • Belgrade

MDPI

Special Issue Editors
John D. Hedengren
Brigham Young University
USA

Logan Beal
Brigham Young University
USA

Editorial Office
MDPI AG
St. Alban-Anlage 66
Basel, Switzerland

This edition is a reprint of the Special Issue published online in the open access journal *Processes* (ISSN 2227-9717) in 2017 (available at: http://www.mdpi.com/journal/processes/special_issues/Combined_Scheduling).

For citation purposes, cite each article independently as indicated on the article page online and as indicated below:

Lastname, F.M.; Lastname, F.M. Article title. *Journal Name*. **Year** *Article number, page range*.

First Edition 2018

ISBN 978-3-03842-805-3 (Pbk)
ISBN 978-3-03842-806-0 (PDF)

Cover photo courtesy of John D. Hedengren and Logan Beal.

Table of Contents

About the Special Issue Editors

John D. Hedengren is Associate Professor at Brigham Young University in the Chemical Engineering Department, leading the PRISM (Process Research and Intelligent System Modeling) group. He is a chemical engineer by training with a B.S. and M.S. degree from Brigham Young University, and a Ph.D. from the University of Texas at Austin. His area of expertise is in process dynamics, control, and optimization with applications in fiber optic monitoring, automation of oil and gas processes, unmanned aerial systems, systems biology, and grid-scale energy systems. He has extensive experience in automation and modeling complex systems. Automation software (APMonitor) he developed has been applied in many industries world-wide including unmanned aircraft systems, chemicals manufacturing, and in energy production.

Logan Beal is a Graduate Research Assistant at Brigham Young University in the Chemical Engineering Department. He leads the PRISM research group on combined scheduling and control and is the key contributor to the National Science Foundation award, "EAGER: Cyber-Manufacturing with Multi-echelon Control and Scheduling". His prior work experience includes optimization research with ExxonMobil and automation at Knoll's Atomic Power Lab. His current research is in developing a computing platform for large-scale dynamic optimization with the GEKKO package for Python. He is also developing a novel approach to nonlinear programming by combining elements of interior point methods and active set Sequential Quadratic Programming (SQP).

Preface to "Combined Scheduling and Control"

Scheduling and control are typically viewed as separate applications because of historical factors such as limited computing resources. Now that algorithms and computing resources have advanced, there are new efforts to have short-term decisions (control) interact or merge with longer-term decisions (scheduling). A new generation of numerical optimization methods are evolving to capture additional benefits and unify the approach to manufacturing process automation.

This special issue is a collection of some of the latest advancements in scheduling and control for both batch and continuous processes. It contains developments with multi-scale problem formulation, software for the new class of problems, and a survey of the strengths and weaknesses of successive levels of integration.

John D. Hedengren and Logan Beal

Special Issue Editors

processes

MDPI

Editorial
Special Issue: Combined Scheduling and Control

John Hedengren * and Logan Beal

Department of Chemical Engineering, Process Research and Intelligent Systems Modeling (PRISM), Brigham Young University, Provo, UT 84602, USA; beall@byu.edu
* Correspondence: john.hedengren@byu.edu; Tel.: +1-801-477-7341

Received: 1 March 2018; Accepted: 2 March 2018; Published: 7 March 2018

This Special Issue (SI) of *Processes*, "Combined Scheduling and Control," includes approaches to formulating combined objective functions, multi-scale approaches to integration, mixed discrete and continuous formulations, estimation of uncertain control and scheduling states, mixed integer and nonlinear programming advances, benchmark development, comparison of centralized and decentralized methods, and software that facilitates the creation of new applications and long-term sustainment of benefits. Contributions acknowledge strengths, weaknesses, and potential further advancements, along with a demonstration of improvement over current industrial best-practice.

Advanced optimization algorithms and increased computational resources are opening new possibilities to integrate control and scheduling. Some of the most popular advanced control methods today were conceptualized decades ago. Over a time span of 30 years, computers have increased in speed by about 17,000 times and algorithms such as integer programming have a speedup of approximately 150,000 times on some benchmark problems. With the combined hardware and software improvements, benchmark problems can now be solved 2.5 billion times faster; i.e., applications that formerly required 120 years to solve are now completed in 5 s [1]. New computing architectures and algorithms advance the frontier of solving larger scale and more complex integrated problems. Recent work demonstrates economic and operational incentives for merging scheduling and control.

The accepted publications cover a range of topics and methods for combining control and scheduling. There were many submissions to the special issue, and about 50% were accepted for publication. The seven that were accepted have novel approaches, summary surveys, and illustrative examples that validate the methods and motivate further investigation. The articles are summarized below.

Lefebvre, D. Dynamical Scheduling and Robust Control in Uncertain Environments with Petri Nets for DESs [2].

This paper is about the incremental computation of control sequences for discrete event systems in uncertain environments through implementation of timed Petri nets. The robustness of the resulting trajectory is also evaluated according to risk probability. A sufficient condition is provided to compute robust trajectories. The proposed results are applicable to a large class of discrete event systems, in particular in the domains of flexible manufacturing.

Joglekar, G. Using Simulation for Scheduling and Rescheduling of Batch Processes [3].

This paper uses a BATCHES simulation model to accurately represent the complex recipes and operating rules typically encountered in batch process manufacturing. By using the advanced capabilities of the simulator (such as modeling assignment decisions, coordination logic, and plant operation rules), very reliable and verifiable schedules can be generated for the underlying process. Scheduling methodologies for a one-segment recipe and a rescheduling methodology for day-to-day decisions are presented.

Gupta, D.; Maravelias, C. A General State-Space Formulation for Online Scheduling [4].

This paper presents a generalized state-space model formulation particularly motivated by an online scheduling perspective, which allows for the modeling of (1) task-delays and unit breakdowns; (2) fractional delays and unit downtimes when using a discrete-time grid; (3) variable batch-sizes; (4) robust scheduling through the use of conservative yield estimates and processing times;

(5) feedback on task-yield estimates before the task finishes; (6) task termination during its execution; (7) post-production storage of material in unit; and (8) unit capacity degradation and maintenance. These proposed generalizations enable a natural way to handle routinely encountered disturbances and a rich set of corresponding counter-decisions, thereby simplifying and extending the possible application of mathematical-programming-based online scheduling solutions to diverse application settings.

Nunes de Barros, F.; Bhaskar, A.; Singh, R. A Validated Model for Design and Evaluation of Control Architectures for a Continuous Tablet Compaction Process [5].

In this work, a dynamic tablet compaction model capable of predicting linear and nonlinear process responses is successfully developed and validated. The applicability of the model for control system design is evaluated and the developed control strategies are implemented on an experimental setup. Evidence that Model Predictive Control (MPC) with an unmeasured disturbance model is the most adequate control algorithm for the studied system is presented. It is concluded that the selection of control strategies for a given compaction process is heavily dependent on real-time measurements of tablet attributes.

Beal, L.; Petersen, D.; Pila, G.; Davis, B.; Warnick, S.; Hedengren, J. Economic Benefit from Progressive Integration of Scheduling and Control for Continuous Chemical Processes [6].

This work summarizes and reviews the evidence for the economic benefit from scheduling and control integration, reactive scheduling with process disturbances, market updates, and a combination of reactive and integrated scheduling and control. This work demonstrates the value of combining scheduling and control and of responding to process disturbances or market updates. The case studies quantify the value of four phases of progressive integration and three scenarios with process disturbances and market fluctuations.

Petersen, D.; Beal, L.; Prestwich, D.; Warnick, S.; Hedengren, J. Combined Noncyclic Scheduling and Advanced Control for Continuous Chemical Processes [7].

This paper introduces a novel formulation for combined scheduling and control of multi-product, continuous chemical processes in which nonlinear model predictive control (NMPC) and noncyclic continuous-time scheduling are efficiently combined. The method uses a decomposition into nonlinear programming (NLP) and mixed-integer linear programming (MILP) problems, an iterative method to determine the number of production slots required, and a filter method to reduce the number of MILP problems required. Results demonstrate the effectiveness and computational feasibility of the approach when dealing with volatile market conditions or a large number of possible products within a short time frame.

Sahlodin, A.; Barton, P. Efficient Control Discretization Based on Turnpike Theory for Dynamic Optimization [8].

In this paper, a new control discretization approach for dynamic optimization of continuous processes is proposed. It builds upon turnpike theory in optimal control and exploits the solution structure for constructing the optimal trajectories and adaptively deciding the locations of the control discretization points. The method is most suitable for continuous systems with sufficiently long time horizons during which steady state is likely to emerge. The proposed adaptive discretization is built directly into the problem formulation, thus requiring only one optimization problem instead of a series of successively refined problems. It is shown that the proposed approach can significantly reduce the computational cost of dynamic optimization for systems of interest.

The papers from this special issue can be accessed at the following link: http://www.mdpi.com/journal/processes/special_issues/Combined_Scheduling.

As this special issue and other recent articles demonstrate, combined scheduling and control is an active area of focus in the process systems engineering community. There also remain several areas for development with optimization algorithms that converge within a controller cycle time, improved scale-up with many discrete variables (especially in MINLP), the exploitation of unique problem structures, and the utilization of strengths with emerging computing architectures. Nonlinear relationships are needed where feedback linearization or linear dynamic models are not sufficient to capture the control dynamics. Further development towards the unification of scheduling and control particularly needs industrial application with guidance on benefits and further development opportunities.

Guest Editors
John Hedengren and Logan Beal
Brigham Young University
Process Research and Intelligent Systems Modeling (PRISM) Group
Brigham Young University
Provo, Utah 84602
USA

References

1. Linderoth, J. *Overview of Mixed-Integer Programming: Recent Advances, and Future Research Directions*; FOCAPO/CPC: Tucson, Arizona, 2017.
2. Lefebvre, D. Dynamical Scheduling and Robust Control in Uncertain Environments with Petri Nets for DESs. *Processes* **2017**, *5*, 54.
3. Joglekar, G. Using Simulation for Scheduling and Rescheduling of Batch Processes. *Processes* **2017**, *5*, 66.
4. Gupta, D.; Maravelias, C.T. A General State-Space Formulation for Online Scheduling. *Processes* **2017**, *5*, 69.
5. Nunes de Barros, F.; Bhaskar, A.; Singh, R. A Validated Model for Design and Evaluation of Control Architectures for a Continuous Tablet Compaction Process. *Processes* **2017**, *5*, 76.
6. Beal, L.D.R.; Petersen, D.; Pila, G.; Davis, B.; Warnick, S.; Hedengren, J.D. Economic Benefit from Progressive Integration of Scheduling and Control for Continuous Chemical Processes. *Processes* **2017**, *5*, 84.
7. Petersen, D.; Beal, L.D.R.; Prestwich, D.; Warnick, S.; Hedengren, J.D. Combined Noncyclic Scheduling and Advanced Control for Continuous Chemical Processes. *Processes* **2017**, *5*, 83.
8. Sahlodin, A.M.; Barton, P.I. Efficient Control Discretization Based on Turnpike Theory for Dynamic Optimization. *Processes* **2017**, *5*, 85.

processes

MDPI

Article

Dynamical Scheduling and Robust Control in Uncertain Environments with Petri Nets for DESs

Dimitri Lefebvre

GREAH Research Group, UNIHAVRE, Normandie University, 76600 Le Havre, France;
dimitri.lefebvre@univ-lehavre.fr

Received: 3 September 2017; Accepted: 21 September 2017; Published: 1 October 2017

Abstract: This paper is about the incremental computation of control sequences for discrete event systems in uncertain environments where uncontrollable events may occur. Timed Petri nets are used for this purpose. The aim is to drive the marking of the net from an initial value to a reference one, in minimal or near-minimal time, by avoiding forbidden markings, deadlocks, and dead branches. The approach is similar to model predictive control with a finite set of control actions. At each step only a small area of the reachability graph is explored: this leads to a reasonable computational complexity. The robustness of the resulting trajectory is also evaluated according to a risk probability. A sufficient condition is provided to compute robust trajectories. The proposed results are applicable to a large class of discrete event systems, in particular in the domains of flexible manufacturing. However, they are also applicable to other domains as communication, computer science, transportation, and traffic as long as the considered systems admit Petri Nets (PNs) models. They are suitable for dynamical deadlock-free scheduling and reconfiguration problems in uncertain environments.

Keywords: discrete event systems; timed Petri nets; stochastic Petri nets; model predictive control; scheduling problems

1. Introduction

The design of controllers that optimize a cost function is an important objective in many control problems, in particular in scheduling problems that aim to allocate a limited number of resources within several users or servers according to the optimization of a given cost function. In the domains of flexible manufacturing, communication, computer science, transportation, and traffic, the makespan is commonly used as an effective cost function because it leads directly to minimal cycle times. However, due to multi-layer resource sharing and routing flexibility of the jobs, scheduling problems are often NP-hard problems. Many recent works in operations research, automatic control, and computer science communities have studied such problems. In operations research community, flow-shop, and job-shop problem have been investigated from a long time [1,2] and a lot of contributions have been proposed, based either on heuristic methods (like Nawaz, Enscore and Ham or Campbell, Dudek, and Smith heuristics) or artificial intelligence and evolutionary theory [3–5]. In the automatic control community, automata, Petri nets (PNs), and max-plus algebra have been used to solve scheduling problems for discrete event systems (DESs) [6,7]. In particular, with PNs, the pioneer contributions for scheduling problems are based on the Dijkstra and A* algorithms [8,9]. Such algorithms explore the reachability graph of the net, in order to generate schedules. Numerous improvements have been proposed: pruning of non-promising branches [10,11], backtracking limitation [12], determination of lower bounds for the makespan [13], best first search with backtracking, and heuristic [14] or dynamic programming [15]. By combining scheduling and supervisory control in the same approach, one can also avoid deadlocks. Some approaches have been proposed: search in the partial reachability graph [16], genetic algorithms [17], and heuristic functions based on the firing vector [13,18]. The performance of operations research approaches are good, in general, compared to the automatic control approaches as long as static scheduling problems are considered. The advantage to solving scheduling problems with PNs or other tools issued from the control theory is to use a common formalism to describe a large class of problems and to facilitate the representation from one problem to another. In particular, PNs are suitable to represent many systems in various domains as flexible manufacturing, communication, computer science, transportation, and traffic [6,7]. This makes such approaches more suitable for dynamic and robust scheduling in uncertain environments. However, modularity and genericity usually suffer from a large computational effort that disqualifies the approaches for numerous large systems.

This work aims to propose a modular and generic approach of weak complexity. It details a method for timed PNs that incrementally computes control sequences in uncertain environments. Uncertainties are assumed to result from system failures or other unexpected events, and robustness with respect to such uncertainties is obtained thanks to a model predictive control (MPC) approach. The computed control sequences aim to reach a reference state from an initial one. The forbidden states, as deadlocks and dead-branches are avoided. The trajectory duration approaches its minimal value. Thanks to its robustness, the proposed approach generates dynamical and reconfigurable schedules. Consequently, it can be used in a real-time context. Resource allocation and operation scheduling for manufacturing systems are considered as the main applications. The robustness of the resulting trajectory is evaluated as a risk belief or probability. For that purpose structural and behavioral models of the uncertainties are considered. Finally, robust trajectories are computed. Compared to our previous works [19–22], the main contributions are: including, explicitly, uncertainties by means of uncontrollable stochastic transitions in the PNs model; evaluating the risk of the computed control sequences; proposing a sufficient condition for the existence of robust trajectories.

The paper is organized as follows. In Section 2, the preliminary notions and the proposed method are developed: timed PNs with uncontrollable transitions are presented, non-robust and robust control sequences are introduced, and the approach to compute non-robust and robust control sequences with minimal duration is developed. Section 3 illustrates the method on a simple example and then presents the performance for a case study. Section 4 is a discussion about the method and the results. Section 5 sums up the conclusions and perspectives.

2. Materials and Methods

2.1. Petri Nets

A PN structure is defined as $G = <P, T, W_{PR}, W_{PO}>$, where $P = \{P_1, \ldots, P_n\}$ is a set of n places and $T = \{T_1, \ldots, T_q\}$ is a set of q transitions with indices $\{1, \ldots, q\}$ $W_{PO} \in (\mathbf{N})^{n \times q}$ and $W_{PR} \in (\mathbf{N})^{n \times q}$ are the post- and pre- incidence matrices ($\mathbf{N}$ is the set of non-negative integer numbers), and $W = W_{PO} - W_{PR} \in (\mathbf{Z})^{n \times q}$ ($\mathbf{Z}$ is the set of positive and negative integer numbers) is the incidence matrix. $<G, M_I>$ is a PN system with initial marking M_I and $M \in (\mathbf{N})^n$ represents the PN marking vector. The enabling degree of transition T_j at marking M is given by $n_j(M)$:

$$n_j(M) = \min\{\lfloor m_k/w^{PR}_{kj} \rfloor : P_k \in {}^{\circ}T_j\} \tag{1}$$

where ${}^{\circ}T_j$ stands for the preset of T_j, m_k is the marking of place P_k, w^{PR}_{kj} is the entry of matrix W_{PR} in row k and column j. A transition T_j is enabled at marking M if and only if (iff) $n_j(M) > 0$, this is denoted as $M [T_j >$. When T_j fires once, the marking varies according to $\Delta M = M' - M = W(:, j)$, where $W(:, j)$ is the column j of incidence matrix. This is denoted by $M [T_j > M'$ or equivalently by $M' = M + W.X_j$ where X_j denotes the firing count vector of transition T_j [7]. A firing sequence σ is defined as $\sigma = T(j_1)T(j_2) \ldots T(j_h)$ where $j_1, \ldots j_h$ are the indices of the transitions. $X(\sigma) \in (\mathbf{N})^q$ is the firing count vector associated to σ, $|\sigma| = ||X(\sigma)||_1 = h$ is the length of σ ($|| \ ||_1$ stands for the 1-norm), and $\sigma = \varepsilon$ stands for the empty sequence. The firing sequence σ fired at M leads to the trajectory (σ, M):

$$(\sigma, M) = M(0) \; [T(j_1) > M(1) \; .... \; M(h-1) \; [T(j_h) > M(h) \tag{2}$$

where $M(0) = M$ is the marking from which the trajectory is issued, $M(1)$, ..., $M(h-1)$ are the intermediate markings and $M(h)$ is the final marking (in the next, we write $M(k) \in (\sigma, M)$, $k = 0, \ldots h$). A marking M is said to be **reachable** from initial marking M_I if there exists a firing sequence σ such that $M_I \; [\sigma > M$ and σ is said to be **feasible** at M_I. $R(G, M_I)$ is the set of all reachable markings from M_I.

2.2. Forbidden, Dangerous and Robust Legal Markings

For control issues, the set of transitions T is divided into two disjoint subsets T_C, and T_{NC} such that $T = T_C \cup T_{NC}$. T_C is the subset of q_C controllable transitions, and T_{NC} the subset of q_{NC} **uncontrollable** transitions. Without loss of generality $T_C = \{T_1, \ldots, T_{qC}\}$ and $T_{NC} = \{T_{qC+1}, \ldots, T_{qC+qNC}\}$. The firings of enabled controllable transitions are enforced or avoided by the controller, whereas the firings of uncontrollable transitions are not, and uncontrollable transitions fire spontaneously according to some unknown random processes. A set of **marking specifications** is also defined with the function *SPEC*: for any marking $M \in R(G, M_I)$, $SPEC(M) = 1$ if M satisfies the marking specifications, otherwise $SPEC(M) = 0$. When no specification is considered, $SPEC(M) = 1$ for all $M \in R(G, M_I)$. The two disjoint sets $F(G, M_I, M_{ref})$ and $L(G, M_I, M_{ref})$ of forbidden and legal markings respectively are introduced:

$$L(G, M_I, M_{ref}) = \{M \in R(G, M_I) \text{ at } \exists \; \sigma \in (T_C)^* \text{ with } M \; [\sigma > M_{ref} \text{ with } (SPEC(M') = 1)$$
$$\text{for all } M' \in (\sigma, M)\} \tag{3}$$

$$F(G, M_I, M_{ref}) = R(G, M_I)/L(G, M_I, M_{ref}) \tag{4}$$

In other words, a marking $M \in R(G, M_I)$ is **legal** with respect to M_{ref} if a trajectory exists from M to M_{ref} that contains only controllable transitions and intermediate markings that satisfy the specifications. In addition, a legal marking M is **robust** with respect to T_C if $M° \subseteq T_C$, where $M°$ stands for the set of transitions enabled at M, otherwise M is **dangerous** (Figure 1) With this definition of robust and dangerous markings, a marking that satisfies $M° \subseteq T_C$ but that has only dangerous markings as successors in $R(G, M_I)$ is considered as robust. Note that a finer partition of the legal markings in three classes (strong robust, weak robust, and dangerous) could be used for some problems. On the contrary, a **forbidden** marking is a marking from which no controllable trajectory exists to the reference. Examples of forbidden markings are deadlocks or markings that do not satisfy the system specifications or markings that enable only uncontrollable transitions (Figure 1).

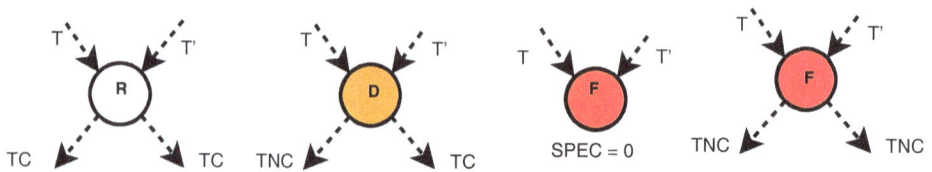

Figure 1. Examples of robust (R), dangerous (D) and forbidden (F) markings in $R(G, M_I)$ depending on the controllable (TC) and uncontrollable transitions (TNC).

The previous definitions are extended to trajectories. A robust trajectory is a legal trajectory that visits only robust markings. On the contrary a dangerous trajectory is a legal trajectory that visits at least one dangerous marking.

2.3. Timed Petri Nets with Uncontrollable Transitions

Timed Petri nets are PNs whose behaviors are constrained by temporal specifications [7]. For this reason, timed PNs have been intensively used to describe DESs like production systems [6]. This paper concerns partially-controlled timed PNs under and infinite server semantic where the firing of controllable transitions behaves according to an earliest firing preselection policy (transitions fire earliest in the order computed by the controller) and time specifications similar to the one used for T-timed PNs [23]: if $T_j \in T_C$, the firing of T_j occurs at earliest after a minimal delay $d_{min\,j}$ from the date it has been enabled ($d_{min\,j} = 0$ if no time specification exists for T_j). On the contrary, the firings of uncontrollable transitions are unpredictable: if $T_j \in T_{NC}$, the firings of T_j occur according to an unknown arbitrarily random process at any time from the date it has been enabled. Consequently, partially-controlled timed PNs (PCont-TPNs) are defined as $<G, M_I, D_{min}>$ where $D_{min} = (d_{min\,j}) \in (\mathbf{R}^+)^{qC}$ and $\mathbf{R}^+$ is the set of non-negative real numbers. If in addition, the stochastic dynamics of the uncontrollable transitions are driven by exponential probability density functions (pdfs) of parameters $\mu = (\mu_j) \in (\mathbf{R}^+)^{qNC}$, with a race policy and a resampling memory [24], then partially controlled stochastic timed PNs (PCont-SPNs) defined as $<G, M_I, D_{min}, \mu>$ will be used instead of PCont-TPNs. The parameters $d_{min\,j}$ are set in an arbitrary time unit (TU) and the parameters μ_j are set in TU^{-1}.

A timed firing sequence σ of length $|\sigma| = h$ and of duration t_h is defined as $\sigma = T(j_1, t_1)T(j_2, t_2) \ldots T(j_h, t_h)$ where $j_1, \ldots j_h$ are the indices of the transitions, and $t_1, \ldots, t_h$ represent the dates of the firings that satisfy $0 \le t_1 \le t_2 \le \ldots \le t_h$. The timed firing sequence σ fired at M leads to the timed trajectory (σ, M):

$$(\sigma, M) = M(0) \, [T(j_1, t_1) > M(1) \ldots . M(h-1) \, [T(j_h, t_h) > M(h) \qquad (5)$$

with $M(0) = M$. Note that, under earliest firing policy, an untimed trajectory of the form of Equation (2) that contains only controllable transitions can be transformed in a straightforward way into a timed trajectory of the form of Equation (5) of minimal duration [20,21] using Algorithm 1. This algorithm also returns $DURATION(\sigma, M) = t_h$.

Algorithm 1. Transformation of an untimed trajectory (σ, M) into timed one (σ', M).

(Inputs: σ, M, G, D_{min}; Output: σ', τ)

1.   initialization: $\tau \leftarrow 0$; $CAL \leftarrow \{(T_j, d_{min\,j})$ at $M\,[\,T_j >\}$, $\sigma' \leftarrow (\varepsilon, 0)$, $h \leftarrow |\sigma|$
2.   for k from 1 to h
3.   find in CAL the date τ_k of the earliest occurrence of the kth transition $T(j_k)$ in σ
4. $\tau \leftarrow \tau_k$, remove entry $(T(j_k), \tau_k)$ in CAL
5. $CAL_{new} \leftarrow \emptyset$, $M' \leftarrow M - W_{PR}.X(T(j_k))$
6.   for all T' at $M'\,[\,T' >$
7.   compute the enabling degree $n'(T', M')$ of T' at M'
8.   for j from 1 to $n(T', M')$
9.   find the jth occurrence (T', τ'_j) of T' in CAL
10. $CAL_{new} \leftarrow CAL_{new} \cup (T', \max(\tau'_j, \tau))$
11.   end for
12.   end for
13. $M'' \leftarrow M' + W_{PO}.X(T(j_k))$
14.   for all t'' at $M''\,[\,T'' >$
15.   compute the enabling degree $n''(T'', M'')$ of T'' at M''
16.   for j from 1 to $n''(T'', M'') - n'(T', M')$
17. $CAL_{new} \leftarrow CAL_{new} \cup (T'', \tau + d_{min}(T''))$
18.   end for
19.   end for
20. $CAL \leftarrow CAL_{new}$, $\sigma' \leftarrow \sigma'\,(T(j_k), \tau_k)$
21.   end for
22. $\tau \leftarrow \tau_h$

2.4. Belief and Probability of Trajectory Deviation

The objective of this section is to evaluate the risk that uncontrollable firings may occur during the execution of the trajectory (σ, M_I) and deviate the trajectory from the reference. For PCont-TPNs, this risk is evaluated with the belief $RB(\sigma, M_I, T_C)$:

$$RB(\sigma, M_I, T_C) = h_{NC}/h \qquad (6)$$

where h_{NC} is the number of intermediate dangerous markings in (σ, M_I) and h is the number of markings visited by (σ, M_I). For PCont-SPNs, the belief $RB(\sigma, M_I, T_C)$ is replaced by the probability $RP(\sigma, M_I, T_C)$ that can be computed with Proposition 1:

Proposition 1. *Let $<G, M_I, D_{min}, \mu>$ be a PCont-SPN, under the earliest firing policy, with M_I a legal robust marking. Let M_{ref} be a reference marking and (σ, M_I) be a legal trajectory to M_{ref}. The probability $RP(\sigma, M_I, T_C)$ that (σ, M_I) deviates from the reference is given by:*

$$RP(\sigma, M_I, T_C) = \sum_{1 \le k_1 \le h} \pi(k_1) - \sum_{1 \le k_1 < k_2 \le h} (\pi(k_1).\pi(k_2)) + \cdots +$$
$$(-1)^{h-1}. \sum_{1 \le k_1 < \ldots < k_{h-1} \le h} (\pi(k_1) \ldots \pi(k_{h-1})) + (-1)^h.\pi(1) \ldots \pi(h) \qquad (7)$$

with:

$$\pi(k) = \frac{\sum_{T_j \in T_{NC} \cup (M(k))^\circ} \mu_j}{\sum_{T_j \in T_{NC} \cup (M(k))^\circ} \mu_j + \left(d_{j_k}\right)^{-1}}$$

if $d_{j_k} \ne 0$, otherwise $\pi(k) = 0$, and $d_{j_k} = t_{k+1} - t_k$ is the remaining time to fire $T(j_{k+1}, t_{k+1})$ at date t_k.

Proof. $RP(\sigma, M_I, T_C)$ is the probability to fire uncontrollable transitions when dangerous markings belong to (σ, M_I).

Consider the trajectory of Figure 2. Under earliest firing policy, the probability that the uncontrollable transition T_{NC1} or T_{NC2} fires before $T(j_{k+1}, t_{k+1})$ and that the trajectory deviates from M_{ref} at $M(k)$ is given by:

$$\pi(k) = Prob(T_{NC1} \text{ or } T_{NC2} \text{ fires before } T(j_{k+1}, t_{k+1})) = \frac{\mu_1 + \mu_2}{\mu_1 + \mu_2 + \left(d_{j_k}\right)^{-1}}$$

if $d_{j_k} \ne 0$, otherwise $Prob(T_{NC1}$ or T_{NC2} fires before $T(j_{k+1}, t_{k+1})) = 0$. Note that if the controllable transition $T(j_{k+1}, t_{k+1})$ fires earliest after a duration d_{j_k}, then the probability $\pi(k)$ is computed by considering the approximation $1/d_{j_k}$ of the mean firing rate of $T(j_{k+1}, t_{k+1})$. Note also that the duration of other controllable transitions enabled at $M(k)$ (for example, T_{C2} in Figure 2) are not considered because this transition does not belong to (σ, M_I). Alternatively the probability that the trajectory continues to $M(k+1)$ at $M(k)$ is given by:

$$1 - \pi(k) = Prob(T(j_{k+1}, t_{k+1}) \text{ fires before } T_{NC1} \text{ and } T_{NC2}) = \frac{\left(d_{j_k}\right)^{-1}}{\mu_1 + \mu_2 + \left(d_{j_k}\right)^{-1}} \qquad (8)$$

Thus, $RP(\sigma, M_I, T_C)$ is finally given by:

$$RP(\sigma, M_I, T_C) = \pi(0) + (1 - \pi(0))(\pi(1) + (1 - \pi(1)) \ldots \pi(h)))$$

for which an exhaustive development is easily rewritten as in Equation (7).

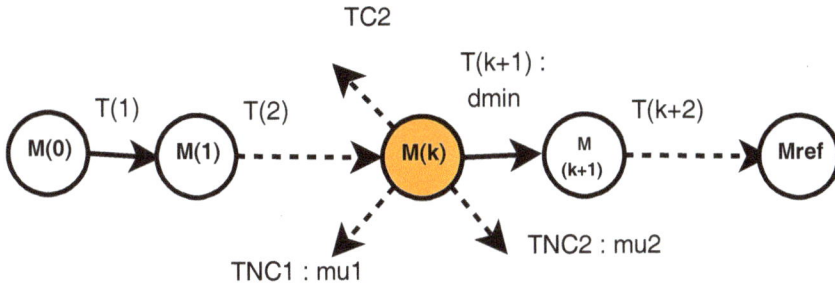

Figure 2. An example of dangerous trajectory: $M(k)$ enables two controllable transitions $T(k + 1)$ and T_{C2} and two uncontrollable ones T_{NC1} and T_{NC2}.

2.5. Model Predictive Control for PCont-TPNs

The determination of control sequences for untimed and timed PNs that contain only controllable transitions has been considered in our previous works [19,20] with a model predictive control (MPC) approach adapted for DESs. In this section, this approach is extended to PCont-TPNs (and consecutively to PCont-SPNs). At each step, the future trajectory is predicted from the current state. A sequence of control actions is computed by minimizing and the first action of the sequence is applied. Then prediction starts again from the new state reached by the system [25,26]. The cost function $J_{FC}(M, M_{ref}) = (D_{min})^T$. X based on the temporal specification and on the evaluation X of the firing count vector, that leads to the reference M_{ref} from the marking M, has been introduced in our previous work [21] to estimate the time to the reference. In this section, this cost function is rewritten for PCont-TPNs. For this purpose let us define G_C and $W_C \in (\mathbf{Z})^{n \times qC}$ as the restrictions of G and W to the set of controllable transitions T_C. The controllable firing count vector X_C that satisfies $M_{ref} - M = W_C.X_C$ and minimizes $J_{FC}(M, M_{ref}) = (D_{min})^T.X_C$ is obtained by solving an optimization problem with integer variables of reduced size $q_C - r$ where r is the rank of W_C. A regular matrix $P_L \in (\mathbf{Z})^{n \times n}$ and a regular permutation matrix $P_R \in \{0,1\}^{qC \times qC}$ exists at:

$$W_C' = P_L.W_C.P_R = \begin{pmatrix} W_{11} & W_{12} \\ W_{21} & W_{22} \end{pmatrix} \tag{9}$$

with $W_{11} \in (\mathbf{Z})^{r \times r}$ a regular upper triangular matrix with integer entries, and $W_{21} = 0_{(n-r) \times r}$, $W_{22} = 0_{(n-r) \times (qC-r)}$ zero matrices of appropriate dimensions. For each $M \in R(G, M_I)$, solving Equation(10):

$$\text{Min } \{(D_{min})^T.X_C : X_C \in (\mathbf{N})^{qC} \text{ at } W_C.X_C = (M_{ref} - M)\} \tag{10}$$

is equivalent to solving Equation (11) and this leads to reduce the number of variables by r:

$$\text{Min } \{F_2.X_{C2} : X_{C2} \in (\mathbf{N})^{qC-r} \text{ at } (W_{11})^{-1}.W_{12}.X_{C2} \leq (W_{11})^{-1}.\Delta M_1\} \tag{11}$$

with $F_2 = (D_{min})^T.(P_{R2} - P_{R1}.(W_{11})^{-1}.W_{12})$, $P_R = (P_{R1} \mid P_{R2})$, $P_L = ((P_{L1})^T \mid (P_{L2})^T)^T$ and $\Delta M_1 = P_{L1}.(M_{ref} - M)$. This reformulation results from the rewriting $(\Delta M_1^T \, \Delta M_2^T)^T = P_L.(M_{ref} - M)$ and $(X_{C1}^T \, X_{C2}^T)^T = (P_R)^{-1}.X_C$ with $X_{C1} = (W_{11})^{-1}.\Delta M_1 - (W_{11})^{-1}.W_{12}.X_{C2}$. The linear optimization problem (Equation (11)) has a solution with integer values as long as $M_{ref} \in R(G_C, M)$ and the cost function $J_{FC}(M, M_{ref})$ based on firing count vector X_{C2} and on D_{min} is defined by Equation (12):

$$J_{FC}(M, M_{ref}) = (D_{min})^T.(P_{R1}.(W_{11})^{-1}.\Delta M_1 + P_{R2}.X_{C2} \, P_{R1}.(W_{11})^{-1}.W_{12}.X_{C2}) \tag{12}$$

As long as X_{C2} corresponds to a feasible and legal firing sequence σ to the reference (i.e., X_{C2} does not encode a spurious solution for Equation (11)), $J_{FC}(M, M_{ref})$ provides an upper bound of the duration of σ as proved with Proposition 2.

Proposition 2. *Let us consider a PCont-TPN (resp. PCont-SPN) of parameter D_{min} (with respect to the parameters D_{min} and μ), under the earliest firing policy. Let M_{ref} be a reference marking and (σ, M_I) a legal trajectory to M_{ref} with $\sigma \in T_C{}^*$ and minimal duration $DURATION(\sigma, M_I)$. Let $X_C(\sigma) \in (N)^{qC}$ be the firing count vector of σ. Then:*

$$DURATION(\sigma, M_I) \leq (D_{min})^T . X_C(\sigma) \tag{13}$$

Proof. (σ, M_I) is written as in Equation (5). $T(j_1, t_1)$ is enabled at date 0 and fires at date $t_1 = d_{min\, j1}$ to result in marking $M(1)$. $T(j_2, t_2)$ is enabled at date 0 or t_1 and fires not later than $t_1 + d_{min\, j2}$. Thus $t_2 \leq d_{min\, j1} + d_{min\, j2}$. The same reasoning is repeated h times. $T(j_h, t_h)$ is enabled at latest at date t_{h-1} and fires not later than $t_{h-1} + d_{min\, jh}$. Thus $t_h \leq d_{min\, j1} + \ldots + d_{min\, jh}$. The minimal duration of (σ, M_I) is t_h, thus, Equation (13) holds.

The basic idea is to use $J_{FC}(M, M_{ref})$ to iteratively drive the search of the controllable firing sequence of minimal duration that leads to the reference. At each step (i.e., for each intermediate marking), a part of the controllable reachability graph is explored and a prediction of the remaining duration to the reference is obtained with cost function $J_{FC}(M, M_{ref})$ computed for each marking M of the explored graph. Then the first control action is applied (i.e., the next controllable transition fires). If an uncontrollable firing occurs, the trajectory deviates from the predicted one and the system enters in an unexpected state. However, the deviation is immediately taken into account by the controller that updates the control sequence at the next step. For this reason the proposed strategy leads to a dynamical and robust scheduling. Two algorithms already developed in our previous works [21,22] are used for that purpose.

Algorithm 2 similar to the one developed in [21,22] encodes as a tree *Tree(M, H)* a small part of the reachability graph rooted at M (Figure 3). The tree is limited in depth with parameter H and in duration with parameter H_τ.

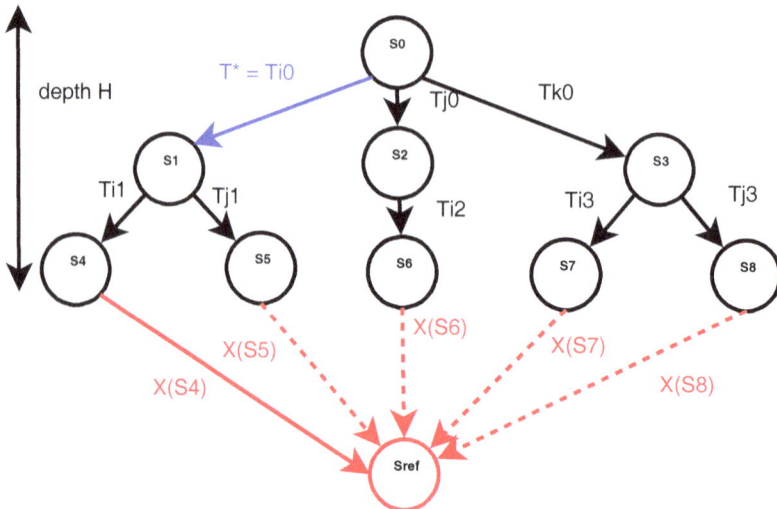

Figure 3. Computation of the next transition to fire with Algorithm 2.

Each node $S = \{m(S), \sigma(S), s(S), l(S), e(S)\} \in Tree(M, H)$ is tagged with a marking $m(S)$, the firing sequence $\sigma(S)$ at M [$\sigma(S) > m(S)$], and the sequence of nodes $s(S)$ in the tree from M to $m(S)$. In addition, the flags $l(S)$ and $e(S)$ are introduced at $l(S) = 0$ if S is forbidden, otherwise $l(S) = 1$ and $e(S) = 1$ if S is a terminal node of the tree, otherwise $e(S) = 0$. At each intermediate marking, Algorithm 2 returns the next transition T^* to fire.

Algorithm 2. Computation of T^* for PCont-TPNs.

(Inputs: $M, M_{ref}, G_C, SPEC, F, H, H_\tau$; Outputs: F, *converge*, *exhaustive*, T^*)

1. if $M \in F, S_0 \leftarrow \{M, \varepsilon, S_0, 0, 1\}$, *converge* $\leftarrow -2$, else $S_0 \leftarrow \{M, \varepsilon, S_0, 1, 0\}$, end if
2. if $M = M_{ref}, S_0 \leftarrow \{M, \varepsilon, S_0, 1, 1\}$, *converge* $\leftarrow 1$,   else $S_0 \leftarrow \{M, \varepsilon, S_0, 1, 0\}$, end if
3. **Tree** $\leftarrow S_0, \Sigma \leftarrow S_0, T^* \leftarrow \varepsilon$, *exhaustive* $\leftarrow 1$
4. while $\exists\, S \in$ **Tree** at $l(S) = 1$ and $e(S) = 0$,
5.   for each $T \in T_C$ at $m(S)$ [$T >$
6.   compute the successor S' of S by firing T, M' at $m(S)$ [$t > M', \sigma' \leftarrow \sigma(S)\, T, s' \leftarrow s(S)\, S'$
7.   if $(SPEC(M') = 0) \vee ((M')^\circ \cup T_C = \varnothing), F \leftarrow F \cup \{m(S)\}$, end if
8.   if $(M' \in F) \vee (S' \in s(S)), l \leftarrow 0$, else $l \leftarrow 1$, end if
9.   if $(l = 0) \vee (M' = M_{ref}) \vee (|\sigma'| = H) \vee (DURATION(\sigma', M) > H_\tau), e \leftarrow 1$, else $e \leftarrow 0$, end if
10.   **Tree** $\leftarrow$ **Tree** $\cup \{M', \sigma', s', l, e\}$
11.   end for
12. end while
13. for h from H-1 to 0
14.   for each $S \in$ **Tree** at $|\sigma(S)| = h$
15.   if $(l(S') = 0$ for all direct successors S' of S in **Tree**), $l(S) \leftarrow 0, e(S) \leftarrow 1$, end if
16.   end for
17. end for
18. for each $S \in$ **Tree** at $(l(S) = 0) \wedge (e(S) = 0)$
19.   if $\exists \perp S' \in$ **Tree** at $(S' \neq S) \wedge (m(S') = m(S)) \wedge (l(S') = 1), F \leftarrow F \cup \{m(S)\}$, end if
20. end for
21. for each $S \in$ **Tree** st $e(S) = 1, \Sigma \leftarrow \Sigma \cup \{S\}$, end if
22. $\Sigma^* \leftarrow \{S^*$ at $J_{FC}(m(S^*), M_{ref}) = \min(J_{FC}(m(S), M_{ref})$, for all $S \in \Sigma\}$
23. $\Sigma^{**} \leftarrow \{S^*$ at $DURATION(\sigma(S^*), M) = \min(DURATION(\sigma(S), M))$ for all $S \in \Sigma^*\}$
24. if $\{S_0\} = \Sigma^{**}$, *converge* $\leftarrow -1, T^* \leftarrow \varepsilon$, else select T^* as the first transition of $\sigma(S^*)$ with $S^* \in \Sigma^{**}$,

    converge $\leftarrow 0$, end if
25. for each $S \in \Sigma$
26.   if $(l(S) = 1) \wedge (e(S) = 1) \wedge (DURATION(\sigma(S), M) < H_\tau)$, *exhaustive* $\leftarrow 0$, end if
27. end for

The complete control sequence σ^* is obtained with Algorithm 3 similar to the one developed in [21,22] that adapts the parameter H in range $[1 : \overline{H}]$ where $\overline{H}$ is an input parameter (Figure 4) that limits the maximal depth of the search in steps. This algorithm starts at initial marking M_I, with no forbidden marking (i.e., $F = \varnothing$) and with minimal depth (i.e., $H = 1$). As long as convergence is ensured, T^* is added to σ^* and the current marking M is updated. Finally Algorithm 3 also evaluates the risk RP of the computed trajectory.

Algorithm 3. Control sequence computation for PCont-TPNs.

(Inputs: M_I, M_{ref}, G, T_C, T_{NC}, SPEC, D_{min}, μ, $\overline{H}$, H_τ ; Outputs: σ^*, success, RP)

1. $M \leftarrow M_I$, converge $\leftarrow 0$, $\sigma^* \leftarrow \varepsilon$, $H \leftarrow 1$, $F \leftarrow \varnothing$, success $\leftarrow 1$
2.   while *(converge < 1)*
3.   compute *converge*, *exhaustive* and $T^* \in T_C$ and update F with Algorithm 2
4.   if *(converge = 0)*^*((exhaustive = 1)* ∨ *((exhaustive = 0)*^*(H = $\overline{H}$)))*,
5.   compute $\sigma^* \leftarrow \sigma^* T^*$ and M at M_I $[\sigma^* > M$
6. $H \leftarrow$ max(1, H-1)
7.   end if
8.   if *((converge = −1)* ∧ *(H = $\overline{H}$))* ∨ *(converge = −2)*,
9.   if *(M ≠ M_I)*,
10.   remove last transition in σ^* and compute M at $M_I[\sigma^* > M$
11.   else
12.   if *(converge = −2)*, *success* $\leftarrow -2$, else *success* $\leftarrow -1$,end if
13.   break
14.   end if
15.   end if
16.   if *((H = $\overline{H}$)* ∧ *(converge = 0)* ∧ *(exhaustive = 0))*, *success* $\leftarrow 0$, end if
17.   if *(H < $\overline{H}$)* ∧ *((converge = −1)* ∨ *((converge = 0)* ∧ *(exhaustive = 0)))*, $H \leftarrow H + 1$, end if
18. end while
19. compute RP with (7)

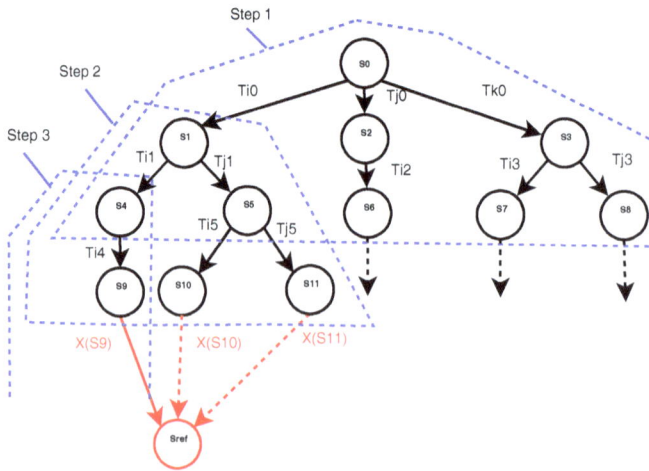

Figure 4. MPC global schema with Algorithm 3.

Note that the complexity of Algorithm 3 is at most $O(h.q_C^{\overline{H}})$ where $h = |\sigma^*|$.

Example 1. *PCont-SPN1 is considered with $T_C = \{T_1, T_2, T_3, T_4, T_5, T_6\}$, $T_{NC} = \{T_7\}$, $D_{min} = (1, 1, 1, 1, 1, 5)^T$ and $\mu = \mu_7 = 1$ (Figure 5). The control objective is to reach $M_{ref} = (5\ 0\ 0\ 0)^T$ from $M_I = (1\ 0\ 0\ 0)^T$ and no additional marking constraint is considered. The cycles $\{P_1, T_1, P_2, T_2\}$ and $\{P_1, T_3, P_3, T_4\}$ are both token producers due to the weighted arcs: the execution of $\{P_1, T_3, P_3, T_4\}$ multiplies each token by 5 compared to $\{P_1, T_1, P_2, T_2\}$ that multiplies it by 2 only. Thus, sequences with cycle $\{P_1, T_3, P_3, T_4\}$ will reach the reference more rapidly. However, the uncontrollable transition T_7 may fire during execution of this cycle which leads to an* excessive production of tokens. The cycle {P_1, T_5, P_4, T_6} which is a token consumer, is then used to correct the excessive number of tokens. Note that the execution of this last cycle is slow compared to the two other ones due (a) to the firing duration of T_6 that is five times larger than the duration of the other transitions; and (b) to the presence of the selfloop {T_5, P_8} that limits the number of simultaneous firings of T_5 to one (whereas the other transitions may fire several times simultaneously according to the infinite server semantic).

The optimal timed sequence to reach M_{ref} is given by $\sigma_1 = T(3, 1)(T(4, 2))^5$ with duration DURATION(σ_1, M_I) = 2 time units (TUs). If no unexpected firing of T_7 occurs, Algorithm 3 applied with $\mathbf{T_C}$ leads to σ_1. However, if unexpected firings of T_7 occur, the trajectory is disturbed and requires more time to reach the reference. Figure 6 is an example of trajectory including one firing of T_7 at date 1.6 TUs. The rest of the control sequence is updated in order to compensate the deviation so that the marking finally reaches M_{ref} in 48.6 TUs instead of 2 TUs.

Figure 6 illustrates the systematic updating of the optimization process at each step (i.e., for each new firing). Consequently the firing of an uncontrollable transition at a given step k changes the future predictions, and the control actions computed at steps $k + 1$, $k + 2$, ... compensate the deviation as long as a controllable trajectory exists from the current marking to M_{ref}.

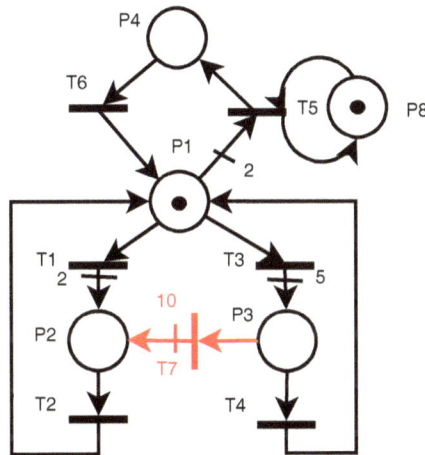

Figure 5. Example PCont-SPN1.

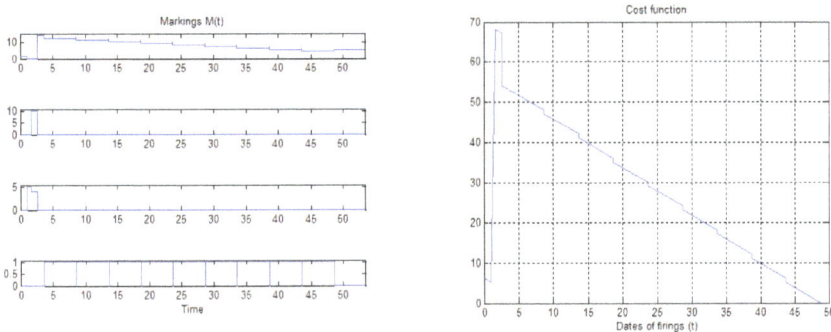

Figure 6. Cost function J_{FC} for a controlled sequence disturbed by an unexpected firing of T_7 with respect to time (TUs).

2.6. Robust Scheduling

In order to compute robust trajectories that cannot deviate from the reference, the controller should avoid dangerous intermediate markings and consider only legal trajectories with robust markings (i.e., with zero-risk belief or probability). The difficulty in this computation is that the intermediate markings are computed step-by-step and these markings are known in advance only within a small time window provided by the part of the reachability graph, of depth H, explored at each step. During the prediction phase of MPC, only the remaining firing count vector to the reference is determined and this vector does not provide the risk belief or risk probability of the future trajectory. Proposition 3 provides a sufficient condition to ensure that the computed trajectory visits only robust markings. For this purpose, let us define $T_{RC} = \{T_j \in T_C \text{ at } (T_j{}^\circ)^\circ \subseteq T_C\}$ where $(T_j{}^\circ)^\circ = \cup \{P_i{}^\circ : P_i \in T_j{}^\circ\}$.

Proposition 3. *Let us consider a Pcont-TPN (or Pcont-SPN). Let (σ, M_I) be a trajectory such that $(M_I)^\circ \subseteq T_C$. If $\sigma \in T_{RC}{}^*$ then (σ, M_I) is a robust legal trajectory.*

Proof. Note at first that $(M_I)^\circ \subseteq T_C$ implies that the net has no uncontrollable source transition (i.e., $^\circ T_j \neq \emptyset$ for all $T_j \in T_{NC}$). Then, (σ, M_I) is written as in Equation (5): $\sigma = M_I [T(j_1, t_1) > M(1) \dots >$ $M(h)$. Assume that there exists $T_j \in (M(1))^\circ$ such that $T_j \in T_{NC}$. T_j is necessarily enabled by the firing of $T(j_1, t_1)$ because T_j is not enabled at M_I. As T_j is not a source transition, there exists a place $P_i \in {}^\circ T_j$ whose marking increases by firing $T(j_1, t_1)$ and consequently $P_i \in (T(j_1, t_1))^\circ$. As $T_j \in P_i{}^\circ$, $T_j \in ((T(j_1, t_1))^\circ)^\circ$. Thus $T_j \in T_C$ that is contradictory with assumption and $(M(1))^\circ \subseteq T_C$. Repeating successively the same reasoning up to $M(h)$, one can conclude that $(M(k))^\circ \subseteq T_C$, $k = 0, \dots, h$, and that (σ, M_I) is a robust legal trajectory. $\blacksquare$

Note that robust legal trajectories are computed with Algorithms 2 and 3 by replacing $W_C \in (\mathbf{Z})^{n \times qC}$ with $W_{RC} \in (\mathbf{Z})^{n \times qRC}$ (i.e., the restriction of W to the set of robust controllable transitions T_{RC}) in the determination of $J_{FC}(M, M_{ref})$.

Note also that the set T_{RC} is easy to obtain by checking for each transition T_j if the condition $X_j.(W_{PO})^T.W_{PR}.(0 \mid I_{qNC})^T = 0$ is satisfied or not, with X_j the firing count vector of T_j and I_{qNC} the identity matrix of size q_{NC}:

$$T_{RC} = \{T_j \in T_C \text{ at } X_j.(W_{PO})^T.W_{PR}.(0 \mid I_{qNC})^T = 0\} \tag{14}$$

Example 2. *Let us consider again Pcont-SPN1 of Figure 5. In order to avoid any deviation, $T_{RC} = \{T_1, T_2, T_4, T_5, T_6\}$ is considered instead of T_C. Algorithm 3 applied with $W_{RC} \in (\mathbf{Z})^{n \times qRC}$ leads to $\sigma_2 = T(1, 1)(T(2, 2))^2(T(1, 3))^2(T(2, 4))^4T(1, 5)(T(2, 6))^2$ that has a duration DURATION(σ_2, M_I) = 6 TUs larger than DURATION(σ_1, M_I).*

The decision to prefer the control sequence σ_2 instead of σ_1 depends on the risk of both control strategies. Table 1 reports the values of RB and RP for both sequences σ_1 and σ_2 with respect to several values of μ. From Table 1, one can notice that the sequence σ_2 that is non-optimal in time has the advantage to be robust compared to σ_1. It cannot be perturbed by any unexpected firing. Note also that the risk probability of σ_1 depends strongly on the dynamic of the random firing of uncontrollable transition T_7. Note finally that computing RP instead of RB provides a better evaluation of that risk.

Table 1. Deviation risk for σ_1 and σ_2.

	RB	RP		
		$\mu_7 = 0.1$	$\mu_7 = 1$	$\mu_7 = 10$
σ_1	$5/6 = 0.83$	$1/3 = 0.33$	$5/6 = 0.83$	$50/51 = 0.98$
σ_2	0	0	0	0

Table 2 reports the mean duration d of control sequences depending on μ_7 for three scenarios. All sequences are computed with $M_I = (1\ 0\ 0\ 0)^T$ and $M_{ref} = (5\ 0\ 0\ 0)^T$ and parameters $\overline{H} = 1$, $H_\tau = 1$. In scenario 1 all transitions are assumed to be controllable. In scenarios 2 and 3, $T_C = \{T_1, T_2, T_3, T_4, T_5, T_6\}$. Algorithm 3 is applied with T_C in scenario 2 whereas it is applied with $T_{RC} = \{T_1, T_2, T_4, T_5, T_6\}$ in scenario 3. Simulations with scenario 2 are repeated 10 times to obtain a significant average duration. One can notice the advantage to compute a robust sub-optimal trajectory with scenario 3, which provides better result from $\mu_7 = 0.5$. When μ_7 increases, the mean duration of T_7 firings decreases and the probability to fire T_7 before T_4 increases; consequently, the number of perturbations increases and the mean duration of the global trajectory also increases due to the execution of the cycle $\{P_1, T_5, P_4, T_6\}$.

Table 2. Performance of Algorithm 3 with Pcont-SPN1: average sequence duration (TUs).

μ_7	0.1	0.5	1	2	10
Scenario 1	2	2	2	2	2
Scenario 2	2.0	20.3	57.2	125.4	228.5
Scenario 3	6	6	6	6	6

3. Results

Pcont-SPN2 (Figure 7) is the timed model of a production system that processes a single type of products according to two possible jobs [27,28]. The first job is composed of the transitions t_1 to t_8, and the second one by the transitions t_9 to t_{14}. In the first job the transitions $T_1, T_3, T_4, T_6, T_7, T_8$ represent the operations in successive machines and the places $P_1, P_2, P_4, P_6, P_7, T_8$ are intermediate buffers where products are temporarily stored. The initial marking of place P_1 represents the maximal number of products that can be simultaneously processed by the Job 1. In the second job the transitions T_9, $T_{10}, T_{11}, T_{12}, T_{13}, T_{14}$ represent the operations in successive machines and the places P_8, P_9, P_{10}, P_{11}, P_{12}, T_{13} are intermediate buffers. The initial marking of place P_8 represents the maximal number of products that can be simultaneously processed by the Job 2. Job 1 could be altered by a server failure whereas Job 2 could not. The occurrence of this failure is represented by the firing of the subsequence $T_2 T_5$ instead of $T_3 T_4$. Note that the faults under consideration are not blocking the system, but they delay the cycle time. Consequently the nominal sequence $T_1 T_3 T_4 T_6 T_7 T_8$ may be altered when an unexpected firing of T_2 occurs that leads to the perturbed behavior $T_1 T_2 T_5 T_6 T_7 T_8$ with an excessive global duration. The six resources p_{14} to p_{19} have limited capacities: $m(p_{14}) = m(p_{15}) = m(p_{16}) = m(p_{17})$ $= m(p_{18}) = m(p_{19}) = 1$. The places p_{20} and p_{21} represent the input and output buffers, respectively, that contain the number of products to be processed either by Job 1 or Job 2. The temporal specifications are given by $D_{min} = (1\ 1\ 2\ 20\ 1\ 1\ 1\ 3\ 3\ 3\ 3\ 3\ 3)^T$ for $T_C = T/\{T_2\}$ and by $\mu_2 = 1$.

Control sequences are computed with $M_I = 3P_1 + 3P_8 + 1P_{14} + 1P_{15} + 1P_{16} + 1P_{17} + 1P_{18} + 1P_{19}$ $+ kP_{20}$ and $M_{ref} = 3P_1 + 3P_8 + 1P_{14} + 1P_{15} + 1P_{16} + 1P_{17} + 1P_{18} + 1P_{19} + kP_{21}$ where k is a varying parameter. The results are reported in Table 3 for $\overline{H} = 5$ and $H_\tau = 20$.

Table 3. Performance of Algorithm 3 with Pcont-SPN2: average sequence duration (TUs).

k	Scenario 1	Scenario 2	Scenario 3
5	45	103.4	72
10	142	213.5	147
15	236	321.9	222
20	325	427.1	297

Another time, three scenarios are considered: in scenario 1 all transitions, including T_2, are assumed to be controllable with $d_{min\ 2} = 1$. In scenario 2, $T_C = T/\{T_2\}$ and Algorithm 3 is applied with T_C. In scenario 3 Algorithm 3 is applied with $T_{RC} = T/\{T_1, T_2\}$. Note, at first, that due to the numerical values of the firing parameters, the cost function prefers Job 1 that has a global duration of 7 TUs to process one product compared to Job 2, which has a global duration of 18 TUs (without considering the constraints due to the limited resources). Thus scenario 1 corresponds to the iterated execution of Job 1. For scenario 2, $\mu_2 = 1$ and $d_{min\,3} = 1$: consequently the probability that an unexpected firing of T_2 occurs is 0.5. When such a firing occurs the long firing duration $d_{min\,5} = 20$ of T_5 compared to $d_{min\,4} = 2$ alters the global duration required to process the product. This explains that scenario 2 leads to longer sequences compared to scenario 1. Scenario 3 is also tested in a stochastic context with the same value of parameters $\mu_2 = 1$ and $d_{min\,3} = 1$. However, the restriction of the control actions in set T_{RC} prefers systematically Job 2 that is robust to the perturbations. Note also that the global duration for $k = 15$ and $k = 20$ is better with scenario 3 than with scenario 2. This is due to the partial exploration of the reachability graph and to the approximation of the remaining sequence duration with cost function J_{FC} that provide solutions with no warranty of optimality.

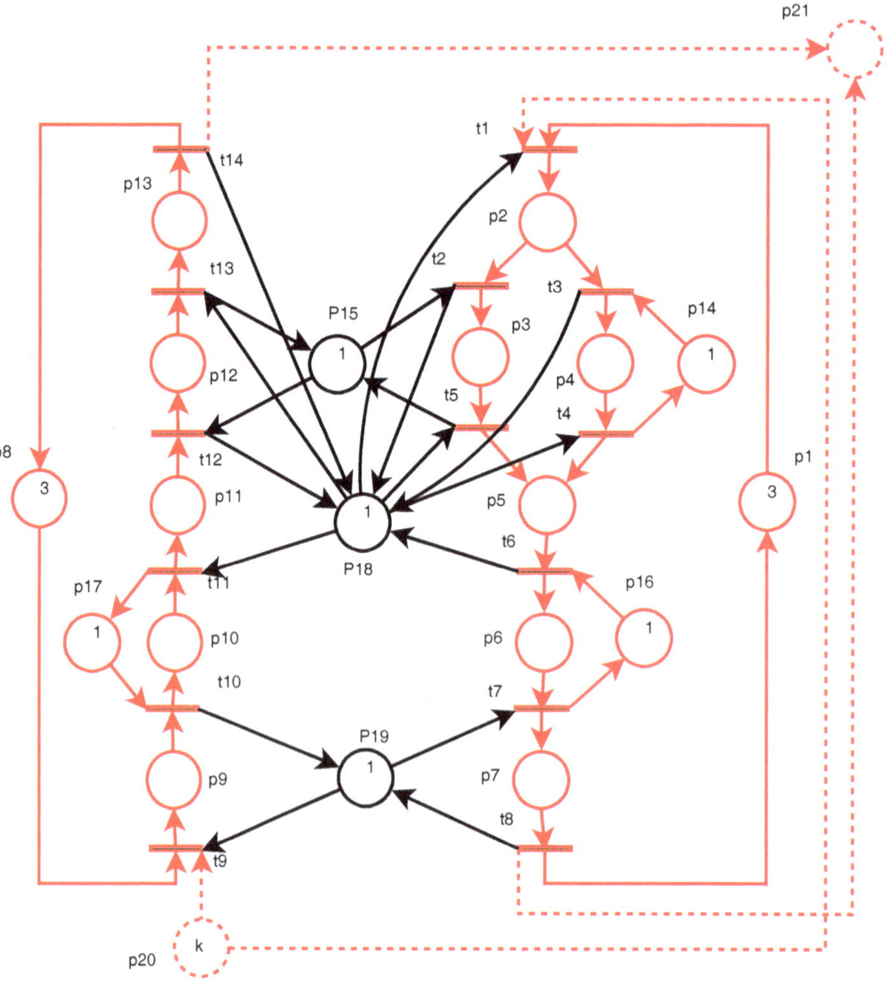

Figure 7. Pcont-SPN2 model of a manufacturing system [28].

4. Discussion

As mentioned in the previous section, the solutions returned by Algorithm 3 are not optimal solutions in a systematic way. The performance of the algorithm depends on the two input parameters: $\overline{H}$, which limits the exploration in depth, and H_τ, which limits the search in duration. If the depth H is too small, Algorithm 2 returns the flag converge = -1 or exhaustive = 0 and Algorithm 3 increases H in the range $[1:\overline{H}]$. On the contrary, if H is too large, then the iterative use of Algorithm 2 certainly reaches M_{ref} but the computational effort is uselessly high. In that case, Algorithm 3 decreases H in the range $[1:\overline{H}]$. Consequently, the aim of Algorithm 3 is to adapt at each step the depth of the search to maintain *converge* = 0 and *exhaustive* = 1 or *converge* = 1. Table 4 reports the performance in function of the parameters $\overline{H}$ and H_τ for Pcont-SPN2 with $M_I = 3P_1 + 3P_8 + 1P_{14} + 1P_{15} + 1P_{16} + 1P_{17} + 1P_{18} + 1P_{19} + 5P_{20}$, $M_{ref} = 3P_1 + 3P_8 + 1P_{14} + 1P_{15} + 1P_{16} + 1P_{17} + 1P_{18} + 1P_{19} + 5P_{21}$, and $T_C = T$. The duration of the control sequences and the computational time required to compute the sequences with Algorithm 3 are reported for an Intel Core i7-46000 CPU at 2.1–2.7 GHz.

Table 4. Performance of Algorithm 3 with respect to parameters $\overline{H}$ and H_τ for PCont-SPN2, sequence duration (TUs) and computational time (s).

$\overline{H}/H_\tau$	1	2	3	4	5	6
5	82 (0.9 s)	86 (0.8 s)	68 (1 s)	68 (1.2 s)	68 (1.3 s)	68 (1.3 s)
10	82 (0.9 s)	86 (0.8 s)	76 (1.5 s)	76 (2.5 s)	76 (4.7 s)	76 (7.9 s)
15	82 (1 s)	86 (0.9 s)	63 (2.1 s)	63 (3.5 s)	63 (9.4 s)	63 (16.1 s)
20	82 (1 s)	86 (0.8 s)	45 (2.6 s)	45 (4.7 s)	45 (10.7 s)	45 (20.6 s)

Note that optimal solutions can be searched in a systematic way instead of using Algorithm 3 considering the extended timed reachability graph [29–31]. Such a graph contains not only the different markings but also the different timed sequences (a given marking can be reached by several sequences with different durations). Table 5 illustrates the rapid increase of the complexity to build such a graph depending on the initial marking $M_I = 3P_1 + 3P_8 + 1P_{14} + 1P_{15} + 1P_{16} + 1P_{17} + 1P_{18} + 1P_{19} + kP_{20}$ when k increases. For each value of k, the number of nodes as the computational time required to compute the graph, are reported for the usual reachability graph and for the timed reachability graph. Table 5 shows that such a method is no longer suitable for large systems. This motivates the proposed approach.

Table 5. Complexity of the exhaustive exploration of control sequences for PCont-SPN2, the number of nodes and the computation time (s).

k	Usual Reachability Graph	Extended Reachability Graph
5	698 (1.4 s)	2208 (106 s)
10	1963 (11 s)	6848 (1827 s)
15	3268 (29 s)	. . .
20	. . .	. . .

5. Conclusions

A method has been proposed to compute control sequences for discrete events systems in uncertain environments. The method uses timed PNs under an earliest firing policy with controllable and uncontrollable transitions as a modeling formalism that is easy to adapt to various problems. The obtained solutions are minimal or near-minimal in duration. Moreover, for each returned solution, the risk to fire uncontrollable transitions is evaluated. Another advantage of the proposed approach is to limit the computational complexity of the algorithm by limiting the part of the reachability graph that is expanded even if the initial marking and reference marking are far from each other, and if deadlocks and dead branches are a priori unknown for the controller. Thanks to the risk evaluation, a robust scheduling becomes computable under some additional assumptions.

In our next works, the research effort will concern, at first, the definition of the cost function that will be improved to provide a more accurate approximation of the remaining time to the reference. The sensitivity of the performance with respect to H will be also studied. We will also include the risk evaluation in the cost function to obtain trajectories of low risk level.

Acknowledgments: The Project MRT MADNESS 2016-2019 has been funded with the support from the European Union with the European Regional Development Fund (ERDF) and from the Regional Council of Normandie.

Conflicts of Interest: The authors declare no conflict of interest.

References

1. Garey, M.R.; Johnson, D.S.; Sethi, R. The complexity of flowshop and jobshop scheduling. *Math. Oper. Res.* **1976**, *1*, 117–129. [CrossRef]
2. Johnson, S.M. Optimal two-and three-stage production schedules with setup times included. *Nav. Res. Logist. Q.* **1954**, *1*, 61–68. [CrossRef]
3. Baker, K.R.; Trietsch, D. *Principles of Sequencing and Scheduling*; John Wiley & Sons: Hoboken, NJ, USA, 2009.
4. Lopez, P.; Roubellat, F. *Production Scheduling*; ISTE: Arlington, VA, USA, 2008.
5. Leung, J.Y. *Handbook of Scheduling: Algorithms, Models, and Performance Analysis*; Chapman & Hall/CRC Computer & Information Science Series: New Delhi, India, 2004; ISBN 9781584883975.
6. Cassandras, C. *Discrete Event Systems: Modeling and Performances Analysis*; Aksen Ass. Inc. Pub.: Homewood, IL, USA, 1993.
7. David, R.; Alla, H. *Petri Nets and Grafcet—Tools for Modelling Discrete Events Systems*; Prentice Hall: London, UK, 1992.
8. Chretienne, P. Timed Petri nets: A solution to the minimum-time-reachability problem between two states of a timed-event-graph. *J. Syst. Softw.* **1986**, *6*, 95–101. [CrossRef]
9. Lee, D.Y.; DiCesare, F. Scheduling flexible manufacturing systems using Petri nets and heuristic search. *IEEE Trans. Robot. Autom.* **1994**, *10*, 123–133. [CrossRef]
10. Sun, T.H.; Cheng, C.W.; Fu, L.C. Petri net based approach to modeling and scheduling for an FMS and a case study. *IEEE Trans. Ind. Electron.* **1994**, *41*, 593–601.
11. Reyes-Moro, A.; Kelleher, H.H.G. Hybrid Heuristic Search for the Scheduling of Flexible Manufacturing Systems Using Petri Nets. *IEEE Trans. Robot. Autom.* **2002**, *18*, 240–245. [CrossRef]
12. Xiong, H.H.; Zhou, M.C. Scheduling of semiconductor test facility via Petri nets and hybrid heuristic search. *IEEE Trans. Semicond. Manuf.* **1998**, *11*, 384–393. [CrossRef]
13. Jeng, M.D.; Chen, S.C. Heuristic search approach using approximate solutions to Petri net state equations for scheduling flexible manufacturing systems. *Int. J. FMS* **1998**, *10*, 139–162.
14. Wang, Q.; Wang, Z. Hybrid Heuristic Search Based on Petri Net for FMS Scheduling. *Energy Proced.* **2012**, *17*, 506–512. [CrossRef]
15. Zhang, W.; Freiheit, T.; Yang, H. Dynamic scheduling in flexible assembly system based on timed Petri nets model. *Robot. Comput. Integr. Manuf.* **2005**, *21*, 550–558. [CrossRef]
16. Hu, H.; Li, Z. Local and global deadlock prevention policies for resource allocation systems using partially generated reachability graphs. *Comput. Ind. Eng.* **2009**, *57*, 1168–1181. [CrossRef]
17. Abdallah, B.; ElMaraghy, H.A.; ElMekkawy, T. Deadlock-free scheduling in flexible manufacturing systems. *Int J. Prod. Res. Vol.* **2002**, *40*, 2733–2756. [CrossRef]
18. Lei, H.; Xing, K.; Han, L.; Xiong, F.; Ge, Z. Deadlock-free scheduling for flexible manufacturing systems using Petri nets and heuristic search. *Comput. Ind. Eng.* **2014**, *72*, 297–305. [CrossRef]
19. Lefebvre, D.; Leclercq, E. Control design for trajectory tracking with untimed Petri nets. *IEEE Trans. Autom. Control* **2015**, *60*, 1921–1926. [CrossRef]
20. Lefebvre, D. Approaching minimal time control sequences for timed Petri nets. *IEEE Trans. Autom. Sci. Eng.* **2016**, *13*, 1215–1221. [CrossRef]

21. Lefebvre, D. Deadlock-free scheduling for Timed Petri Net models combined with MPC and backtracking. In Proceedings of the IEEE WODES 2016, Invited Session "Control, Observation, Estimation and Diagnosis with Timed PNs", Xi'an, China, 30 May–1 June 2016; pp. 466–471.
22. Lefebvre, D. Deadlock-free scheduling for flexible manufacturing systems using untimed Petri nets and model predictive control. In Proceedings of the IFAC—MIM, Invited Session "DES for Manufacturing Systems", Troyes, France, 28–30 June 2016.
23. Ramchandani, C. Analysis of Asynchronous Concurrent Systems by Timed Petri Nets. Ph.D. Thesis, MIT, Cambridge, MA, USA, 1973.
24. Molloy, M.K. Performance analysis using stochastic Petri nets. *IEEE Trans. Comput. C* **1982**, *31*, 913–917. [CrossRef]
25. Richalet, J.; Rault, A.; Testud, J.; Papon, J. Model predictive heuristic control: Applications to industrial processes. *Automatica* **1978**, *14*, 413–428. [CrossRef]
26. Camacho, E.; Bordons, A. *Model Predictive Control*; Springer: London, UK, 2007.
27. Uzam, M. An optimal deadlock prevention policy for flexible manufacturing systems using Petri net models with resources and the theory of regions. *Int. J. Adv. Manuf. Technol.* **2002**, *19*, 192–208. [CrossRef]
28. Chen, Y.; Li, Z.; Khalgui, M.; Mosbahi, O. Design of a Maximally Permissive Liveness-Enforcing Petri Net Supervisor for Flexible Manufacturing Systems. *IEEE Trans. Aut. Science and Eng.* **2011**, *8*, 374–393. [CrossRef]
29. Berthomieu, B.; Vernadat, F. State Class Constructions for Branching Analysis of Time Petri Nets. In Proceedings of the Ninth International Conference on Tools and Algorithms for the Construction and Analysis of Systems TACAS 2003, Warsaw, Poland, 7–11 April 2003; Springer: New York, NY, USA, 2003; Volume 2619, pp. 442–457.
30. Gardey, G.; Roux, O.H.; Roux, O.F. Using Zone Graph Method for Computing the State Space of a Time Petri Net. In Proceedings of the International Conference on Formal Modeling and Analysis of Timed Systems FORMATS 2003, Marseille, France, 6–7 September 2003; Springer: Berlin/Heidelberg, Germany, 2003; Volume 2791, pp. 246–259.
31. Klai, K.; Aber, N.; Petrucci, L. A New Approach to Abstract Reachability State Space of Time Petri Nets. In Proceedings of the 20th International Symposium on Temporal Representation and Reasoning, Pensacola, FL, USA, 26–28 September 2013.

processes

MDPI

Article

Using Simulation for Scheduling and Rescheduling of Batch Processes

Girish Joglekar

Batch Process Technologies, Inc., 112 Eden Court, West Lafayette, IN 47906, USA; girish53@gmail.com

Received: 3 September 2017; Accepted: 26 October 2017; Published: 2 November 2017

Abstract: The problem of scheduling multiproduct and multipurpose batch processes has been studied for more than 30 years using math programming and heuristics. In most formulations, the manufacturing recipes are represented by simplified models using state task network (STN) or resource task network (RTN), transfers of materials are assumed to be instantaneous, constraints due to shared utilities are often ignored, and scheduling horizons are kept small due to the limits on the problem size that can be handled by the solvers. These limitations often result in schedules that are not actionable. A simulation model, on the other hand, can represent a manufacturing recipe to the smallest level of detail. In addition, a simulator can provide a variety of built-in capabilities that model the assignment decisions, coordination logic and plant operation rules. The simulation based schedules are more realistic, verifiable, easy to adapt for changing plant conditions and can be generated in a short period of time. An easy-to-use simulator based framework can be developed to support scheduling decisions made by operations personnel. In this paper, first the complexities of batch recipes and operations are discussed, followed by examples of using the BATCHES simulator for off-line scheduling studies and for day-to-day scheduling.

Keywords: batch process; scheduling; simulation; coordination control; rescheduling

1. Introduction

The problem of scheduling multiproduct and multipurpose batch processes has been studied extensively over the past 30 years [1]. Scheduling involves making decisions for the assignment of tasks to processing units, and the sequencing of various products through the processing facility. Typically, each product is made according to its unique recipe. In some methodologies, the underlying recipes are modeled using either a State-task network (STN) or resource-task network (RTN), and a mixed integer linear programming (MILP) formulation based on discrete time or continuous time representation is used for solving the optimization problem. In some problems where orders can be treated as individual batches, each order moves through various production stages as a discrete entity, and a sequential MIP formulation with order-indexed decision variables for assignments to equipment units and precedence decisions are used to solve the problem. In both the approaches, the manufacturing recipes are greatly simplified in order to keep the problem size to a manageable level so that it can be solved in a reasonable amount of time.

Some of the assumptions made in simplifying the recipes can have significant impact on the solutions generated from these formulations. For example, consider the simple STN shown in Figure 1. It consists of three states, S1, S2 and S3, two tasks T1 and T2, and two units U1 and U2, one for each task. The processing times of the two tasks are t1 and t2, respectively.

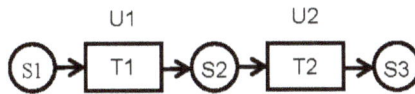

Figure 1. A simple state-task.

In a typical MILP formulation, material transfers S1 to T1, T1 to S2, S2 to T2 and T2 to S3 would be assumed to be instantaneous. The resulting schedule for making two batches of T1 is shown as a Gantt chart in Figure 2. Since the transfers are instantaneous, the end of T1 matches the start of T2.

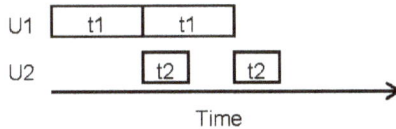

Figure 2. Schedule with zero transfer times.

However, in a real process the transfers are rarely instantaneous. The actual recipe of the two operations may be closer to the recipe network shown in Figure 3. As shown, each task may consist of 3 subtasks, Fill, Mix and Empty. Suppose that processing times t1 and t2 defined earlier are the durations of the Mix subtasks of the two tasks. The transfer of material into task T1 takes place during its Fill subtask, and the transfer from task T1 into T2 takes place during their Empty and Fill subtasks, respectively. As a result, both units, U1 and U2, must be active simultaneously during the transfer step. Suppose the recipe network in Figure 3 is used to drive a simulation model. As shown in the Gantt chart generated by the simulator, shown in Figure 4, the first batch on U2 starts at time f1 + t1, where f1 is the duration of the Fill subtask of task T1 (shown as solid black rectangle). Similarly, the start time of the second batch of T1 will be at (f1 + t1 + e1), where e1 is the duration of the Empty subtask of T1. The duration of the Fill subtask of T2 is the same as that of the Empty subtask of T1, namely e1. The duration of the Empty subtask of T2 is e2, shown as green rectangle. Therefore, the start times of the batches on U2 as predicted by the MILP solution would not be the same as the actual process. It is immaterial how the durations of the material transfers are accounted for, whether they are included in the task times t1 and t2, or treated as additions to t1 and t2. This simple example illustrates that regardless of how the task durations are interpreted, the 'schedule' on the actual process will not match the solution generated by the scheduler when transfer times are not ignored. As the transfer times become more significant, the reliability of the scheduler reduces further.

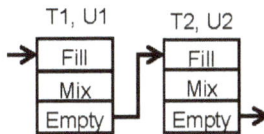

Figure 3. A simple recipe network.

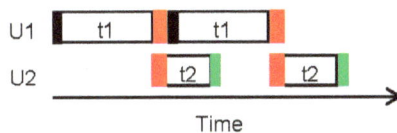

Figure 4. Schedule with non-zero transfer times.

Another important characteristic of batch processes is that the underlying recipes are inherently very complex structurally as well as logistically. For example, consider the recipe network shown in Figure 5 for the manufacture of product P1.

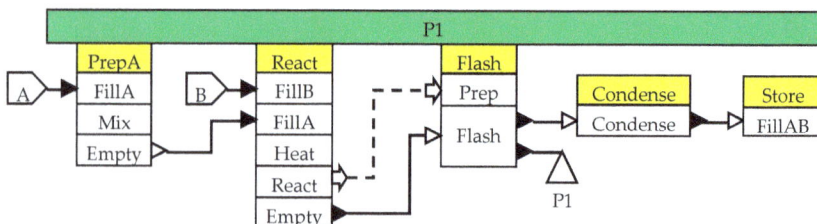

Figure 5. Recipe for manufacture of product P1.

A task in a recipe may consist of several subtasks which are executed in the specified sequence once that task starts in the assigned piece of equipment. For example, task React (yellow in color) in Figure 5 consists of 5 subtasks (white boxes) FillB, FillA, Heat, React and Empty. The flow of material between tasks takes place during specific subtasks. Thus, material from task PrepA is transferred into task React only when the pieces of equipment assigned to them are in Empty and FillA subtasks respectively. Some subtasks form a semi-continuous chain, for example, subtask Flash in Figure 5. The Flash subtask takes material from the Empty subtask of the React task, splits it into two streams, sending one stream to the Condense subtask, which in turn sends it to the FillAB subtask, while the second stream is removed as product P1. Thus, in order to execute the Flash subtask, units assigned to tasks React, Flash, Condense and Store must be in the appropriate subtasks. In MILP based formulations, subtask level details are rarely considered at their fullest level, and semi-continuous transfers across multiple tasks are often treated as single tasks.

In batch processes, the resource usage is typically at the subtask level. For example, the Heat subtask of the React task may require Low Pressure Steam, the React subtask may require Cooling Water, the Flash subtask may require High Pressure Steam, the Condense subtask may require Refrigeration, and so on.

The logistics of coordinating various operations so that the subtasks associated with material transfers are ready just in time are at the heart of scheduling complex recipe driven batch processes. Also, if resources are required in order to execute subtasks, and if there are constraints on the maximum availability of resources, then the resolution of the competition for resources becomes part of the logistics.

Several other factors, such as process variability, operating rules, multiple suitable equipment, equipment dependent batch size and cycle times, sequence dependent cleaning/setup, and so on, influence the operation of a batch process. If the underlying recipes are simplified for solving a scheduling problem, which is often the case in MILP based formulations, the resulting solutions are more likely to be 'broad brush' solutions than 'actionable' solutions. An 'actionable' solution is the information the operations personnel can use with certainty.

The paper by Joglekar [2] showed that using simplified recipe models that are identical to the ones used in MILP formulations, a simulator can generate high quality schedules in a fraction of time as compared to MILP based techniques. This paper demonstrates a simulation based methodology, which incorporates all the complexities of batch processes described above, for generating schedules that are very realistic or 'actionable', and verifiable. Simulation has not been extensively used in addressing scheduling problems, mainly because of the inherently myopic view that this technique takes in making assignment decisions. Simulation coupled with heuristics has been applied to discrete systems. However, in general the characteristics of batch processes are very complex compared to discrete systems. The complexities preclude the use of discrete event simulation based systems in applications related to batch processes4. Simulation based framework applied to batch processes, as reported by Chu et al. [3] augments the underlying MP based formulations, and has similar limitations as stated earlier. The paper by Petrides et al. [4] discusses the roles played by simulators and finite capacity scheduling tools in optimizing biopharmaceutical batch processes, but the tools employed lack the ability to accurately model the underlying complexities.

Often, it is necessary to reschedule a batch process because of changes in external factors used in generating the original schedule, and due to the deviations in the predicted versus actual process trajectories. The rescheduling or reactive scheduling techniques often use RTN based MILP formulations [5,6] and therefore have the same limitations as discussed earlier. The simulation based methodology presented in this paper lends itself very easily to continuous monitoring and rescheduling. Although simulation based approach does not guarantee optimality, performance criteria based on equipment and resource utilization can be established to evaluate the quality of the results. Moreover, simulation predicts time series data for selected process variables and generates subtask based Gantt charts which can be used for tracking the underlying process over time. A supervisory control system could trigger rescheduling decisions based on deviations in the process trajectory.

The following are some of the commercially available simulators for batch processes: Batch Process Developer [7], SuperPro Designer and SchedulePro [8], gPROMS [9], BATCHES [10]. DynoChem [11] is designed for simulating single unit operations used in batch processes. None of these simulators, other than BATCHES, has the ability to dynamically assign a task to a piece of equipment, a key functionality required for solving scheduling problem. In addition, several discrete event simulators are available for discrete manufacturing systems. However, in general batch processes are very complex compared to discrete systems, which precludes the use of discrete simulators for applications related to batch processes.

In this study, the basic methodology of simulation based scheduling is illustrated with a simple process using the BATCHES simulator. First, the various aspects of modeling recipes and the time advance mechanism of the simulator are discussed. Next, the iterative process of doing 'what ifs ... ' to explore the parameter space is explained. At the end, the solution of a scheduling problem is presented.

2. Recipe Models

The need for a systematic framework for defining and managing process recipes was felt strongly by the process control and automation groups within the batch process industry. The basic concepts and terminology for batch process control were adopted in 1995 as the ANSI/ISA-88.01 standard (S88), with addition of more parts later, and an update in 2010 [12,13]. Over the past 20+ years, the S88 standards have been used extensively by the automation vendors. The manufacturing recipe information is at the core of the three kinds of control for batch processes: basic, procedural and coordination [14], and represents the ultimate level of detail that is needed to run the underlying process. A manufacturing recipe is typically implemented by a distributed control system (DCS), such as DeltaV. For example, an operation named REACT in a biopharmaceutical process consists of the following 13 phases [15]: Setup, Add A, Select Path, Receive Product, Dissolve, Add B, Adjust pH, Reaction, Hold, Transfer, CIP Setup, Rinse, and Caustic Wash. The categories of information associated with each phase are: Description, Formula Parameters, Report Parameters, Run Logic, Hold Logic, Abort Logic, Failure Conditions and Stop Logic. In addition, there is documentation of Alarm conditions, Interlocks and shutdowns. For this example, the functional specifications of just one unit procedure spanned 75 pages, with most of the information stored in the DCS system. Programming and maintaining a DCS require significant resources during commissioning and operation of a process. Since a simulation model is a surrogate of the process, if the same level of detail as the control recipe is incorporated into simulated recipes, the simulated results would be very accurate.

In a BATCHES simulation model, a recipe is represented by a top level graphical network like the one shown in Figure 5. The basic building blocks of a recipe network are: Recipe, Task, Subtask,

Raw material (pentagon), Sink (vertical triangle), material input (hollow or solid triangle on the left vertical edge of a subtask), material output (hollow or solid triangle on the right vertical edge of a subtask), flow line (solid line connecting material output to input), signal start (hollow arrow on the right vertical edge of a subtask), signal end (hollow arrow on the left vertical edge of a subtask), signal (dotted line connecting signal start and end). A recipe is simply a name given to a set of tasks. A task in a recipe network defines the set and sequence of elementary steps performed in the assigned unit, and is similar to a unit procedure in S88. A subtask is an elementary step that represents a specific physical/chemical change and is similar to a phase in S88. A sequence of subtasks defines a task, like a series of phases defines a procedure. A recipe network is like a general recipe in S88. A simulation model also consists of an equipment network which defines the set of units at a specific site. Each unit is a physical resource that is required for performing a task. Associated with each task is a list of equipment items that are suitable to perform that task. Thus, a specific combination of recipe and equipment network defines a site recipe in S88, and a specific batch of a task in a specific piece of equipment defines a master recipe. A batch is one instance of execution of a task on a unit. Typically, a recipe produces one material output stream that is considered primary and may produce additional streams that are considered secondary. A batch of material is the quantity of the primary stream produced when the associated task is executed on one of the suitable units. The size of each batch may or may not be the same, dictated by the combination of the recipe and the unit on which that task was performed. In order to produce the required amount of the primary stream, multiple batches are typically made.

The parameters associated with tasks and subtasks define the details of the associated recipe[6]. The important task parameters are: list of suitable equipment items, logic that determines how a piece of equipment is assigned to the task. The important subtask parameters are: dynamic model that best describes the physical/chemical changes taking place during that subtask, how a subtask ends, operator requirements, utility requirements, and state event conditions. The parameters associated with flow lines determine the amount transferred and the associated flowrate.

3. Model and Process Execution

From the discussion in the previous section, it is clear that a BATCHES recipe network can accurately represent the details of a product's manufacturing recipe, and is structurally very similar to the information that drives the batch control software. The main difference between simulation and procedural control is that in order to run a process through time the simulator has to make the key decisions of assigning units to tasks at different points in time, whereas the control software typically relies on an operator to make that decision. An operator, in turn, makes that decision based either on a predefined schedule or on the current process status and experiential knowledge. Once a task is initiated on a unit both the simulator and procedural control implement the recipes. Thus the assignment decisions made by the simulator during a simulation run generate the schedule, which can be given to the operators or the procedural control system as the blue print for running the process. In this section, the key concepts used by the simulator to make the assignment are discussed.

3.1. Time Advance Mechanism in BATCHES

Time advance mechanism is the algorithm used by the simulator to march the specified process through time. The BATCHES simulator uses the time advance mechanism common in combined discrete and dynamic simulators [16,17]. In addition to the recipe driven batch and semi-continuous processes, the simulator also can be used effectively for discrete manufacturing systems. At any discontinuity (event), the simulation executive tries to assign a unit to a task or assign the required material and resources to start a subtask. After all assignments are completed at an event, the simulator integrates in time the state variables associated with all active subtask models until the next event occurs (time or state event). The following are the key decisions that determine the outcome of a simulation run: the assignment of a unit to initiate a task, the assignment of resources to start a subtask and its associated flows, end a subtask and advance the unit to the next subtask, after ending the last subtask of a task end the task and make the associated unit available for next assignment. These decisions predict the process trajectory under the specified conditions, which in essence if the resulting schedule.

3.2. Subtask Types

The type associated with a subtask plays a role in the unit and resource assignment decisions made by the simulator, and is determined by its graphical representation in the recipe network. Since the graphical representation is a modeling decision, a recipe network is a visualization of various subtask level interactions and the decision logic. There are four subtask types: Master, Slave, Chaining and Decoupling. The graphical coding of the subtask types is shown in Figure 6 [18].

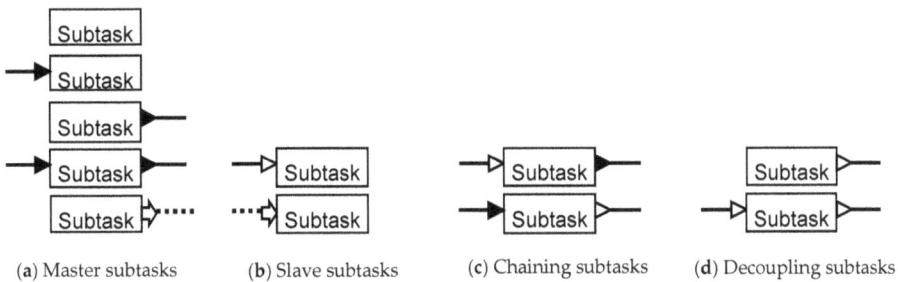

(a) Master subtasks   (b) Slave subtasks   (c) Chaining subtasks   (d) Decoupling subtasks

Figure 6. Graphical depictions of subtask types.

More precise information is needed to determine how flows are executed in a batch process. A material input to a subtask can be a hollow or solid triangle on its left vertical edge. A material input that pulls material from upstream is depicted as a solid triangle. If a material input accepts material pushed by an upstream output it is depicted as a hollow triangle, a passive input.

A material output from a subtask can be a hollow or solid triangle on its right vertical edge. A material output that pushes material downstream is depicted as a solid triangle. If a material output allows downstream input to pull material then it is depicted as a hollow triangle, a passive output.

Thus, the output and input on a material transfer line must be of opposite 'polarity', that is, either the output pushes material (solid triangle) and input receives material (hollow triangle) or the input pulls material (solid triangle) and the output allows material withdrawal (hollow triangle).

Based on its graphical representation, the following rules determine a subtask's type.

Master subtask: A subtask is a master subtask if it has

- no material inputs or output and no signal in or out, or
- only pulling input(s) and no material output(s), or
- only pushing output(s) and no material input(s), or
- pulling input(s) and pushing output(s), or
- only signal start(s).

The various depictions that make a subtask a master subtask are shown in Figure 6a.

Slave subtask: A subtask is a slave subtask if it has a passive input and no material output, or has a signal end and no material input or output. The various depictions that make a subtask a slave subtask are shown in Figure 6b.

Chaining subtask: A subtask is a chaining subtask if it has a passive input and a pushing output, or has a passive output and a pulling input. In the first case, the upstream subtask pushes material into a chaining subtask, and in turn the chaining subtask pushes material downstream (forward chaining). In the second case, the downstream subtask pulls material from a chaining subtask, and in turn the chaining subtask pulls material from upstream (backward chaining). The two depictions that make a subtask a chaining subtask are shown in Figure 6c. A chaining subtask is part of a semi-continuous chain.

Decoupling subtask: A subtask is a decoupling subtask if it has a passive output and no material input, or a passive input and a passive output. A decoupling subtask allows the downstream subtask to pull material from it, and if it has passive input allows the upstream subtask to push material into it asynchronously. The depictions that make a subtask a decoupling subtask are shown in Figure 6d. A storage tank is a typical example of a decoupling subtask.

A subtask can have any number of material inputs and outputs, each connected to a different upstream or downstream subtask. All inputs 2 and above are solid triangles, that is, they pull upstream material. All outputs 2 and above are solid, that is, they push material downstream.

In a batch process, a semi-continuous chain rarely forms a single or nested recycle loop. If there is a continuous recycle loop, then exactly one subtask in the loop must be a decoupling subtask. A recipe network may form a batch recycle loop where material leaving a task during one subtask may return to the same task during another subtask, which would necessarily be at a different time.

When a task advances into a subtask, the following are the main steps in executing that subtask: wait until all the necessary conditions to start the subtask are satisfied, start all the subtasks controlled by the master subtask, implement the actions associated with the subtask as specified by the user, detect when the conditions for ending the subtask are satisfied, end the subtask, inform all other interacting subtask and advance the task to the next subtask. When the last subtask of a task is completed, the task is ended and the equipment assigned to it is released.

3.3. Task Types

The type associated with a task plays a role in the assignment of a unit to that task. A task can be independent or dependent. A task is independent if its first subtask is of type master. Otherwise, it is a dependent task. Like subtasks, task type is determined by its graphical representation in the recipe network.

An independent task can be initiated through a sequence directive. In its simplest form, a sequence directive identifies the task to be initiated, the suggested start time for the first batch of the task, the number of batches of the task to be initiated, and the minimum time elapsed between consecutive initiations of the task (the intra-entry time). Note that the two time parameters define lower bounds. As described in the time advance mechanism, the simulation executive can only check if a task can be assigned to a unit at any discontinuity, it cannot guarantee assignment at a particular time. That is determined by the state of the process at that time. Similarly, the actual time elapsed between two consecutive batches of a task may be greater than or equal to the specified intra-entry time.

A dependent task is initiated when an upstream unit advances into a subtask that pushes material into its first subtask. The upstream subtask generates a request, and the queue processing mechanism initiates a task and the material is transferred downstream.

3.4. Recipe Network Topology

Based on the structure of the graphical model of a recipe, several patterns can be identified that can be used to guide the simulator in making assignment decisions.

A recipe can be entirely front-end driven. Consider the recipe network shown in Figure 7a. It has only one independent task, T1.

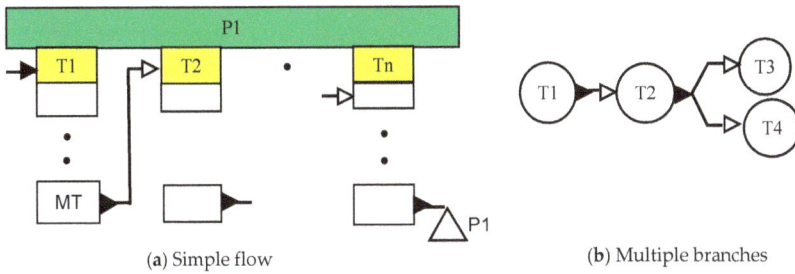

(a) Simple flow   (b) Multiple branches

Figure 7. Front-end driven recipe patterns with single independent task.

Once a batch of T1 is initiated, say through a processing directive, task T2 is started when a unit in task T1 reaches the MT subtask. Thus, T2 is 'scheduled' by T1. The same applies to all other tasks in a front-end driven recipe. A different pattern, shown in Figure 7b, follows the same principle that an upstream subtask triggers a downstream task but has multiple branches that are triggered. Same or different subtasks of T2 may trigger T3 and T4. Of course, these patterns could be arbitrarily complex at the subtask level.

The recipe pattern shown at the task level in Figure 8 can also be considered as front-end driven, but has multiple independent tasks.

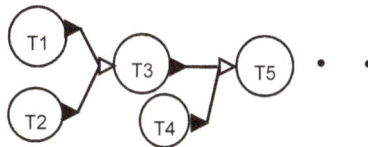

Figure 8. Front-end driven recipe patterns with multiple independent task.

Tasks T1, T2 and T4 are independent in this example presented in Figure 8.

Another common pattern in recipes consists of front-end driven segments that are joined by decoupling subtasks, such as a storage tank, as shown in Figure 9.

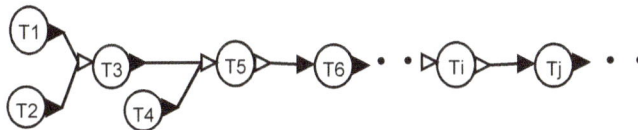

Figure 9. Front-end driven recipe patterns with decoupling subtask.

Tasks T5 and Ti are storage operations. Tasks upstream of T5 form a front-end driven segment, tasks between T5 and Ti form another front-end driven segment starting with T6, and the tasks downstream of Ti form a third front-end driven segment. Typically, the materials in tasks T5, Ti etc. are stable intermediates that can be stored for a long time. Additionally, each segment may represent a facility that is at different physical location.

In addition, various combinations of the front-end driven segments with or without intermediate storage are possible, including recycle of material. Some examples are shown in Figure 10.

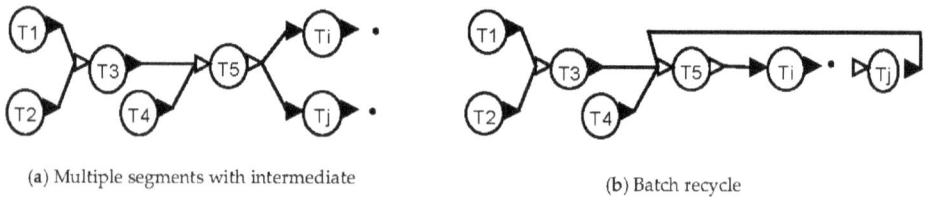

(a) Multiple segments with intermediate

(b) Batch recycle

Figure 10. Additional front-end driven recipe patterns with decoupling subtask.

In Figure 10a, multiple segments use the intermediate from storage task T5. This pattern would represent a process where multiple finished products are made from an intermediate. In Figure 10b, material from Tj is recycled to T5. This would represent a process where a chemical is recovered, such as a solvent, and recycled in a batch recycle mode.

Typically, the most important decision variable associated with each front-end driven segment of a recipe is the time between the successive batches of an independent task in that segment.

4. Simulation Based Scheduling Methodology

In this section, a user directed iterative approach to using simulation for generating a schedule is presented. The basic steps are shown in Figure 11.

The first step is to build a simulation model that has the appropriate level of detail. As a general rule, the more detailed and accurate the model of underlying process, the better the results predicted by the simulator. The model building phase often serves as a way to capture knowhow from various stakeholders and incorporate all significant operational level details into the model, the most important component being the recipe networks associated with the various products.

The next step is to develop a process signature for each recipe. This is achieved by initiating enough batches of the independent tasks to produce some amount of finished product(s). Typically initiating one batch each of independent tasks is sufficient to develop a preliminary process signature. If there are constraints on the process, it may be necessary to run the model for longer time so that all the constraints are manifested during the run. Developing process signature may require an exploratory approach where process complexities are added gradually, providing insights into process behavior at each stage. The following are some of the parameters that broadly characterize a process: effective batch cycle time, batch size of each stage, average throughput of each stage and lag times between any two subtasks in a recipe. At 'steady state', the average throughput of each stage is equal to the throughput of the slowest stage (the bottleneck).

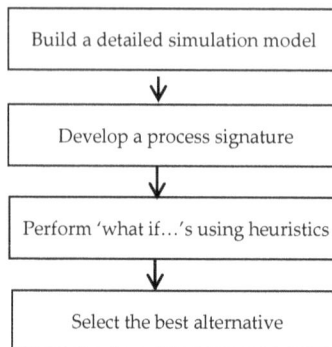

Figure 11. Steps in simulation based scheduling methodology.

The process signature is crucial in specifying the information that drives a simulation run. One of the driving mechanisms used for triggering independent tasks is 'sequence directive'. A sequence directive specifies the time at which to start the first batch, elapsed time between successive batches, and number of batches to initiate of the specified independent task. The guidelines described below for specifying the sequence directives constitute the heuristics used in running the process:

—   For an independent task, the time lag between successive batches in a stage depends of the effective batch cycle time and the number of parallel units in that stage

—   If a stage has multiple independent tasks, the task to start first can be identified from the recipe and the offset for the triggering of the first batch of other independent tasks can be determined from the process signature

During a typical simulation study, several simulation runs are made by changing the values of the desired input parameters. In this iterative approach, the results of a run are analyzed, and the values to be tried next are determined based on the analysis, trends established from prior runs and experiential knowledge.

5. Scheduling of a Specialty Chemical Process

The methodology of using simulation for scheduling is illustrated with a specialty chemical process. The recipe network used in this example is shown in Figure 12.

The manufacture of product P1 consists of 6 operations (tasks in yellow rectangles) REACT, PREPA, BFILTER, FLASH, COLDWCOND and SOLST. The chemical reaction between A and B takes place in a solvent SOL and in the presence of a catalyst CAT. After the reaction, the catalyst is separated using a batch filter. The filtrate is transferred to a batch flash unit where heat is added and the vapor is condensed in a condenser, and the condensate is stored in a solvent storage tank. Most of the solvent and reactants are recovered during the flash, and the remaining liquid is removed as the product P1. Some product is entrained in the condensate. For this study, the recipe was simplified to accumulate the recovered solvent. Therefore, there is only one subtask in the SOLST task. In the actual process, the recovered solvent would be recycled into the REACT and PREPA tasks. The key recipe details are given in Table 1.

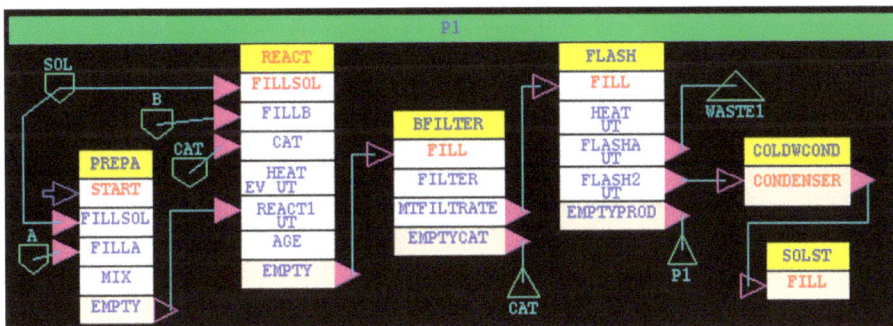

Figure 12. Recipe network of a specialty chemical process.

Table 1. Details of the recipe for P1.

Task	Subtask	Duration (h)	Material	Utilities
REACT	FILLSOL	2.0	SOL, 760 kg	
	FILLB	1.0	B, 136.24 kg	
	CAT	1.0	CAT, 24 kg	
	HEAT	3.33		STEAM, 50 MJ/h
	REACT1	5		COLD WATER, 40 MJ/h
	AGE	0.5		
	EMPTY	0.5	To FILL	
PREPA	START	0.0		
	FILLSOL	1.0	SOL, 760 kg	
	FILLA	0.25	A, 30 kg	
	MIX	1.0		
	EMPTY	0.0	To REACT1	
BFILTER	FILL	0.5	From EMPTY	
	FILTER	0.427		
	MTFILTRATE	0.5	To FILL	
	EMPTYCAT	0.5	To Sink P1	
FLASH	FILL	0.5	From MTFILTRATE	
	HEAT	0.5		STEAM, 100 MJ/h
	FLASHA	0.1		STEAM, 100 MJ/h
	FLASH2	8.2		STEAM, 100 MJ/h,
COLDWCOND	CONDENSER	8.2	From FLASH2 To FILL	COLDWATER, 100 MJ/h

The recipe has one independent task, REACT. The PREPA task is initiated by a signal (signal input on subtask START) triggered 1.0 h after the HEAT subtask of the REACT task starts. Thus, the remainder of HEAT (2.33 h) is enough to cover the 2.25 h required for the first 3 subtasks of PREPA task, allowing the material to be just in time for feeding the reactor.

Two utilities, STEAM and COLDWATER, are required at the specified rates of consumption for the entire duration of the associated subtask. Each utility is constrained at the maximum rate of plant-wide consumption of 100 MJ/h.

One piece of equipment is suitable to perform each task.

The objective of the study is to determine the maximum amount of P1 that can be produced over a span of 168 h assuming that the plant is empty at the beginning.

The simulation runs were made on an HP 15 TouchSmart laptop, IntelCore i3-3110M@2.64 GHz processor, running the Xubuntu 16.04 64-bit operating system.

5.1. Process Signature

To get the process signature, one batch of the REACT task was initiated.

The time required to process the material completely is 24.06 h, and the slowest task is REACT with batch cycle time of 13.33 h, which is also the process cycle time. The Gantt chart for this run is shown in Figure 13.

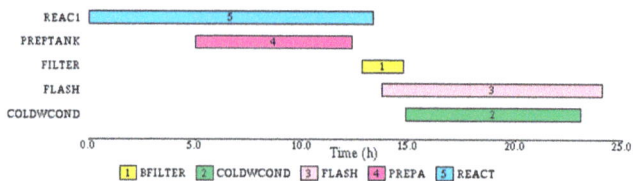

Figure 13. Gantt chart for the one-batch run.

5.2. Base Case

Since the cycle time is 13.33 h, for the given time horizon of 168 h it should be possible to make 12 complete batches. Accordingly, for this run the sequence directive was set to initiate 12 batches of the REACT task without any delay between successive batches.

The simulation shows that the time required to make 12 batches is 233.7 h, much higher than expected. The reason for the increased makespan is the constraint on the STEAM utility. When HEAT, FLASHA and FLASH2 subtasks are running in FLASH, all of the available STEAM utility is consumed. Therefore, even if the reactor batches are started without any delay, the HEAT subtask of REACT task has to wait for STEAM to become available. In the status based Gantt chart for REAC1, shown in Figure 14a, the waiting times are shown in red. The subtask based Gantt chart in Figure 14b shows that the waiting occurs during the HEAT subtask. The waiting time on REAC1 can be eliminated by forcing idle time between successive batches on the reactor so that the end of FLASH2 is synchronized with the beginning of subtask HEAT in the reactor. This delay is approximately 19.06 h. Thus, the effective cycle time of the process is 19.06 h, thereby reducing the number of batches that can be made in the specified time.

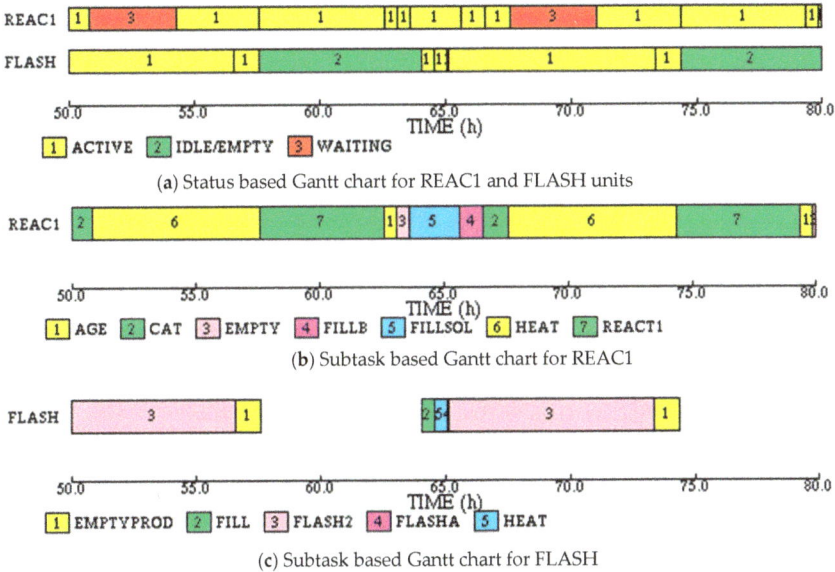

(a) Status based Gantt chart for REAC1 and FLASH units

(b) Subtask based Gantt chart for REAC1

(c) Subtask based Gantt chart for FLASH

Figure 14. Subtask and status based Gantt charts for REAC1 and FLASH for base case.

Another simulation run was made with 19.06 h elapsed time between the starts of successive reactor batches. This eliminates the waiting times on the reactor. The makespan is still 233.7 h. The task based Gantt chart for this run is shown in Figure 15.

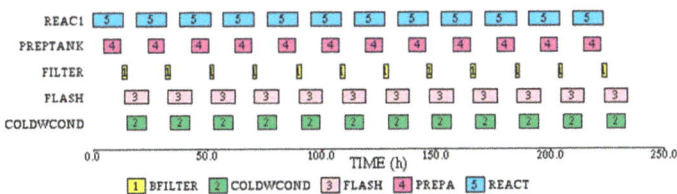

Figure 15. Gantt chart with 19.06 h elapsed time between reactor batches.

For this case 8 batches can be completely processed in 168 h. This case also represents the maximum number of batches that can be made under the given constraints. Thus, due to the constraints the process cycle time increases from 13.33 to 19.06 h, a 43% increase over the unconstrained case.

5.3. Accurate Heat Consumption Profile

It should be noted that the timing of the competition for STEAM by two different subtasks creates a bottleneck, and also their relative positions in the recipe structure cannot be changed. A detailed study of the FLASH operation showed that the demand for STEAM reduces towards the end of FLASH2 subtask. This presents an opportunity to reduce the delays between successive REACT batches if the constraints are not violated. The consumption profile predicted by the dynamic model was approximated by the stepwise function given in Table 2.

A simulation run was made based on the new STEAM consumption profile and delay of 16.7 h between reactor batches. The Gantt chart for this case is shown in Figure 16.

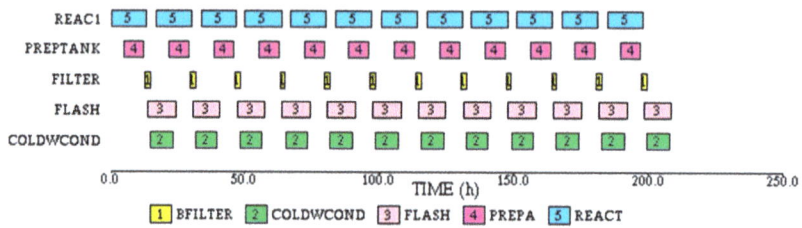

Figure 16. Gantt chart for the case with modified utility usage.

The STEAM consumption profile for the base case is shown in Figure 17a, and current case is shown in Figure 17b.

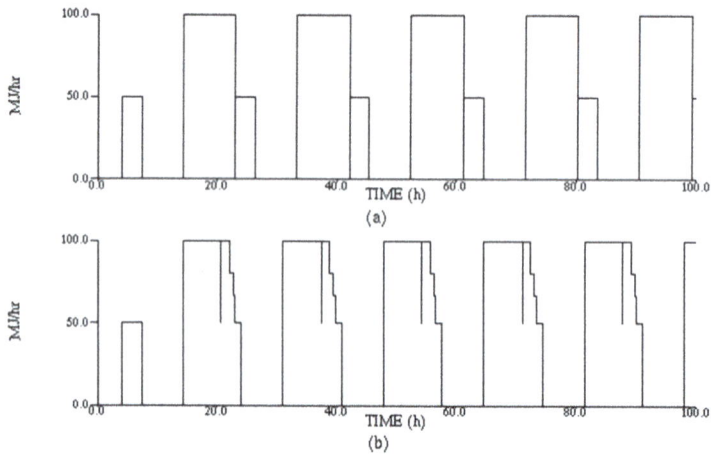

Figure 17. Steam consumption profiles for two cases: (**a**) the base case, (**b**) the current case.

The effective cycle time for this case is reduced to 16.7 h. Consequently, 9 reactor batches can be completely processed in 168 h, which is also the maximum possible throughput of the process. Thus, a more accurate modeling of the utility consumption reduces the effective cycle time or increases the throughput by 12.4%.

Table 2. Consumption profile for STEAM in FLASH2 subtask.

Task	Subtask	Duration (h)	Material	Utilities
FLASH	FLASH2	8.2		STEAM, 100 MJ/h, 5.9 h STEAM, 50 MJ/h, 1.3 h STEAM, 30 MJ/h, 0.7 h STEAM, 17 MJ/h, 0.3 h

6. Use of Simulation in Rescheduling

A recipe network in a BATCHES simulation model very accurately represents the corresponding manufacturing recipe in an automation system used for running the process. Therefore, the results of a simulation run provide accurate and detailed information about the projected process milestones. For example, a subtask level Gantt chart, like the one shown in Figure 14, can be generated for each unit, defining the start and end of each subtask in that unit. A subtask Gantt chart provides a template that could be used for tracking the underlying process in real-time. Similarly, if detailed process dynamics models are used for any subtask, the predicted process variable trajectories also could be used for tracking the process performance. More importantly, checkpoints can be set up based on elapsed time or process state for computing process deviations. At a selected checkpoint, if the difference between the projected value and the actual value of the specified variables is too large, a new schedule could be generated.

The main causes of deviations between a simulated and a real process are process variability and inaccuracies in the process dynamics models. Process variabilities occur due to the randomness in the underlying phenomena, and are typically beyond control or inherent to the process. The inaccuracies in dynamic models arise due to the assumptions made during model formulations. When using simulation as an off-line tool for decision support, multiple simulation runs are made (replicates) with randomized parameters and average values for performance measures are computed. When using a simulation model as a process surrogate in real-time applications, such as rescheduling, fixed values for parameters are used for predicting the process trajectory, with the initial state of the model at each rescheduling point set to match the process state at that time.

The steps in using simulation for rescheduling are shown in Figure 18. At the beginning, the state of each unit is set to match the process state at t = 0. Typically at the beginning each unit is idle and empty.

Most of the automation systems have the functionality to generate a snapshot of the process at any given time. For each unit, the following information can be extracted from the process historian: current subtask, time at which the current subtask started, amount and composition of the material in the unit, temperature and pressure, names of units connected upstream and downstream at current time.

A process snapshot, which can be saved as a text file, is used to set up the initial state of all units in the model and the simulation is halted at a pre-determined end time (time increment b0), for example, 24 h after the start time. The results from the simulation run are made available in the desired format for tracking purposes. As the process marches in time, the status at any time t can be extracted and compared with that projected by the simulation. If the deviation is greater than the specified criteria, then the initial status of the model is set to the process snapshot at that time, and a new simulation run is made. The results of the new simulation are then used for tracking the process from that time forward.

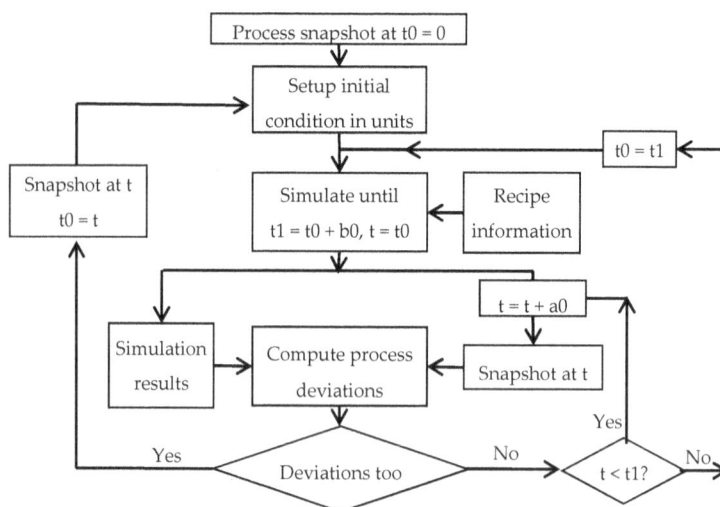

Figure 18. Steps in simulation based rescheduling.

Rescheduling of the Specialty Chemical Process

Simulation based rescheduling is illustrated with the specialty chemical process described earlier. Also, the recipe was modified to include more operational level details that make the model more realistic, and the need for rigorous methodology even more evident.

The HEAT subtasks of the REACT and FLASH tasks require more time for each batch because of fouling. The increases are 0.2 h and 0.1 h, respectively. After cleaning the associated units for 2.0 h after every third batch, the durations for these subtasks are equal to the nominal durations in Table 1. Also, the duration of the REACT1 subtask has variability, and is uniformly distributed between 4 and 6 h.

The variability in the REACT1 subtask and the drift in the HEAT subtask of the REACT task affect the start times of the FILTER and FLASH tasks. Similarly, the drift in the HEAT subtask of the FLASH task further affects the remaining subtasks of the task. Due to the resource constraint, there is a cascading effect on the following REACT batch. Therefore, a schedule generated by assuming fixed cycle time will not be very reliable and there is a need to develop a rescheduling strategy using the most up to date process status.

Based on the knowledge of the process recipe, the end of the REACT1 subtask of the REACT task is used at a check point for rescheduling the process. From practical standpoint, the AGE, EMPTY and FILTER subtasks together provide a time window for making a simulation run, and generating and disseminating the new schedule to the operations personnel.

The rescheduling strategy is as follows:

(1) Start the process at time 0.
(2) Take a process snapshot when the REACT1 subtask of the REACT task ends and note the current time.
(3) In the simulation model, initialize the units to match the snapshot, set the simulation start time to the current time, run the simulation until the REAC1 subtask of the next REACT batch ends.
(4) Trigger the REACT task 1.9 h after the start of FLASH2 subtask of the FLASH task so that the end the first segment of Steam use matches the beginning of HEAT.
(5) Generate the necessary reports for the operations personnel.
(6) Let the process run its course to the end of REACT1 subtask of the next REACT batch.
(7) Go to step 2.

Step 4 given above illustrates the modeling of coordination control in simulation, which is triggering of a task through the use of signals.

The recipe was modified to model the new features. The cleaning operations are modeled as separate tasks, RCLEAN for cleaning the REACT and FLCLEAN for cleaning FLASH, each with one subtask with the duration of 2.0 h. The RCLEAN task is triggered after every 3rd batch on REACT, and FLCLEAN is triggered after every 3rd batch on FLASH. The triggering was implemented through a special purpose FORTRAN subroutine. The modified recipe network is shown in Appendix B. Note that in the simulation used for predicting the process behavior, the duration of the REACT1 subtask is 5.0 h. Therefore, the end of REAC1 predicted by simulation will not match exactly with the end of REACT1 in the process because of the variability. The effect of variability is minimized by adjusting the start time for the simulation run for every cycle and regenerating the schedule one cycle at a time.

The composite Gantt chart for the actual process, simulated by sampling duration of the REACT1 subtask, is shown in Figure 19.

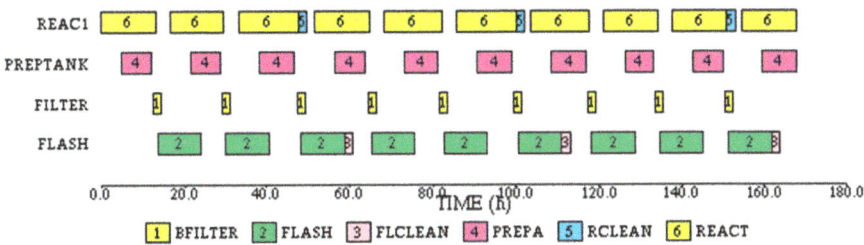

Figure 19. Composite Gantt chart for the specialty chemical process.

The subtask Gantt chart for REAC1 is shown in Figure 20. Note that the durations of the first three instances of HEAT subtask in REAC1 and FLASH tasks are different because of the drift, and durations of all instances of REACT1 subtask are different because of variability.

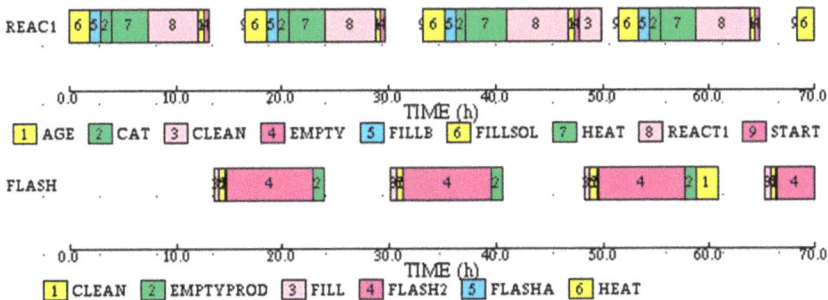

Figure 20. Subtask Gantt chart for REAC1 and FLASH for the specialty chemical process.

For all the simulation runs presented in this paper the execution times were very small, of the order of 100 milliseconds.

7. Conclusions

A recipe network in a BATCHES simulation model can accurately represent the complex recipes and operating rules typically encountered in batch process manufacturing. Based on the modeling of flow and execution controls, a recipe network can be divided into segments that are decoupled from operational standpoint. Each segment can be scheduled independently, assuming no net accumulation in each segment. By using the advanced capabilities of the simulator for making assignment decisions, very reliable and verifiable schedules can be generated for the underlying process. For a recipe consisting of one segment, the best schedule was generated by manipulating one decision variable, namely, the time interval between successive batches of the independent task in the recipe. This methodology has the advantage of keeping the number of decision variables very small, which will facilitate its use in a two level optimization framework.

A rescheduling methodology for making day-to-day scheduling decisions using the simulator was presented for a recipe consisting of one segment. It generates schedules for shorter time horizons and is applied successively. It compensates for the variability in the process through the use of coordination controls incorporated in the recipe model. Such a methodology is feasible because the model can be initialized using a process snapshot generated by a DCS system, and the execution times for making simulation runs are very small.

Conflicts of Interest: The author is associated with Batch Process Technologies, Inc., which develops and licenses the BATCHES simulator. The simulator was used in modeling the processes studied in this paper.

Appendix A

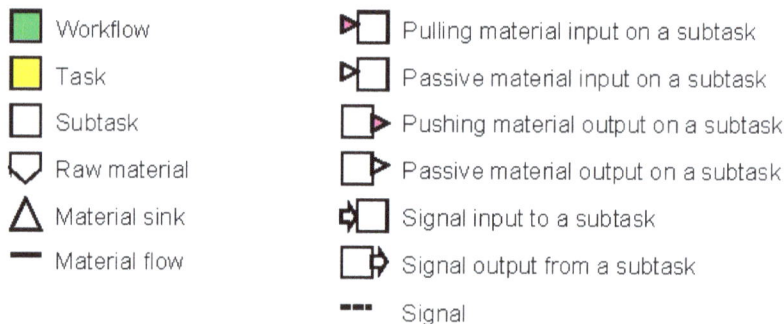

Figure A1. Recipe network Symbols.

Appendix B

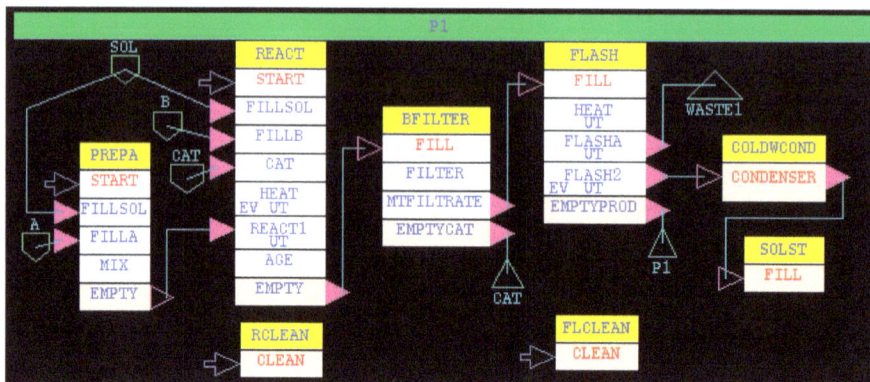

Figure A2. Recipe network for the modified recipe.

References

1. Sundaramoorthy, A.; Maravelias, C.T. A General Framework for Process Scheduling. *AIChE J.* **2011**, *57*, 695–710. [CrossRef]
2. Joglekar, G.S. Incorporating Enhanced Decision-Making Capabilities into a Hybrid Simulator for Scheduling of Batch Processes. *Processes* **2016**, *4*, 30. [CrossRef]
3. Chu, Y.; You, F.; Wassick, J.M.; Agarwal, A. Integrated planning and scheduling under production uncertainties: Bi-level model formulation and hybrid solution method. *Comput. Chem. Eng.* **2015**, *72*, 255–272. [CrossRef]
4. Petrides, D.; Carmichael, D.; Siletti, C.; Koulouris, A. Biopharmaceutical Process Optimization with Simulation and Scheduling Tools. *Bioengineering* **2014**, *1*, 154–187. [CrossRef] [PubMed]
5. Nie, Y.; Biegler, L.T.; Wassick, J.M. Extended Discrete-Time Resource Task Network Formulation for the Reactive Scheduling of a Mixed Batch/Continuous Process. *Ind. Eng. Chem. Res.* **2014**, *53*, 17112–17123. [CrossRef]
6. Gupta, D.; Maravelias, C.T.; Wassick, J.M. From rescheduling to online scheduling. *Chem. Eng. Res. Des.* **2016**, *116*, 83–97. [CrossRef]
7. Batch Process Developer. Available online: www.aspentech.com/products/engineering/aspen-batch-process-developer (accessed on 15 August 2017).
8. Intelligen, Inc. Available online: www.intelligen.com (accessed on 15 August 2017).
9. gPROMS. Available online: www.psenterprise.com/gproms.html (accessed on 15 August 2017).
10. BATCHES. Available online: www.bptechs.com (accessed on 15 August 2017).
11. DynoChem. Available online: www.scale-up.com (accessed on 15 August 2017).
12. ANSI/ISA–88.00.01-2010. *Batch Control Part 1: Models and Terminology*; Instrument Society of America: Pittsburgh, PA, USA, 2010.
13. De Minicis, M.; Giordano, F.; Poli, F.; Schiraldi, M.M. Recipe Development Process Re-Design with ANSI/ISA-88 Batch Control Standard in the Pharmaceutical Industry. *Int. J. Eng. Bus. Manag.* **2014**, *6*, 16–27. [CrossRef]
14. ISAPublications InTech Magazine. Available online: https://www.isa.org/standards-and-publications/isa-publications/intech-magazine/2012/june/automation-basics-organizing-batch-process-control (accessed on 15 August 2017).
15. Watson, D.R.; Eli Lilly and Company, Indianapolis, IN, USA. Personal commumication, 2017.
16. BATCHES. *Users Manual*; Batch Process Technologies: West Lafayette, IN, USA, 2015.
17. Joglekar, G.S.; Reklaitis, G.V. A Simulator for Batch and Semi-continuous Processes. *Comput. Chem. Eng.* **1987**, *8*, 315–327. [CrossRef]
18. Joglekar, G.S.; Giridhar, A.; Reklaitis, G.V. A Workflow Modeling System for capturing data provenance. *Comput. Chem. Eng.* **2014**, *67*, 148–158. [CrossRef]

processes

MDPI

Article

A General State-Space Formulation for Online Scheduling

Dhruv Gupta and Christos T. Maravelias *

Department of Chemical and Biological Engineering, University of Wisconsin-Madison,
Madison, WI 53706, USA; dgupta6@wisc.edu
* Correspondence: christos.maravelias@wisc.edu; Tel.: +1-608-265-9026

Received: 2 October 2017; Accepted: 2 November 2017; Published: 8 November 2017

Abstract: We present a generalized state-space model formulation particularly motivated by an online scheduling perspective, which allows modeling (1) task-delays and unit breakdowns; (2) fractional delays and unit downtimes, when using discrete-time grid; (3) variable batch-sizes; (4) robust scheduling through the use of conservative yield estimates and processing times; (5) feedback on task-yield estimates before the task finishes; (6) task termination during its execution; (7) post-production storage of material in unit; and (8) unit capacity degradation and maintenance. Through these proposed generalizations, we enable a natural way to handle routinely encountered disturbances and a rich set of corresponding counter-decisions. Thereby, greatly simplifying and extending the possible application of mathematical programming based online scheduling solutions to diverse application settings. Finally, we demonstrate the effectiveness of this model on a case study from the field of bio-manufacturing.

Keywords: state-space model; uncertainty; mixed-integer linear programming; model predictive control; bio-manufacturing

---

1. Introduction

Scheduling plays an important role in all industrial production facilities [1]. Contingent on the scale of operation, optimization based scheduling methods can even achieve multi-million dollars increase in profits [2]. Thus, considerable effort has been devoted towards developing optimization models that accurately represent the decision making *flexibility* in these facilities [3]. Maravelias (2012) [4] provides a unified notation and a systematic framework for the description of chemical scheduling problems. Further, significant advances in solution methods, now enable us to solve small size scheduling problems. For example, a highly constrained scheduling instance over a network of 8 processing units, 19 tasks, and 26 materials, with a realistic scheduling horizon of 2 weeks, was shown to be solved to optimality in less than 1 min on an ordinary office computer [5]. Thus, being able to generate and revise schedules in an online fashion, so as to account for new information and disturbances, is very much a reality now. The Dow Chemical Company has already adopted online scheduling in many of its production facilities [6–8].

Scheduling models, as have been developed till now, were not necessarily designed with an emphasis on being natively ready for implementation in an online scheduling setting. Thus, the online framework utilizing a model had to be tailored to that specific model, and required many ad-hoc (heuristic) adjustments to be able to represent and resolve a disturbance to the schedule [9–23]. The introduction of the state-space *idea* to chemical production scheduling alleviated many of these issues that arose from having to make ad-hoc adjustments [24].

In this work, we present a generalized and extended state-space model which is suitable for implementation in an online scheduling setting. Further, we propose a new scheme for updating the state of the process, as well as an overall formulation to enforce constraints (through parameter/variable modifications), based on feedback information, on future decisions. Although here, we focus on expanding the modeling scope, it is important to point out that once a model has been adopted, there are still many other factors which influence the performance of an online scheduling method [25]. These factors are: the online optimization horizon length, the re-computation trigger and its frequency if periodic, allowable changes from one online iteration to the next, any added constraints (e.g., terminal constraints), and the modeling of uncertainty (deterministic vs. stochastic optimization) [8].

This paper is structured as follows. In Section 2, we present a brief background on chemical production scheduling and discuss the state-space model of Subramanian et al. [24]. In Section 3, we present a reformulated state-space model, based on a new convention, and showcase the generalizations on it one at a time. In Section 4, we present the final integrated model, with all generalizations present simultaneously, which requires more than simply concatenating all the individual generalizations together. Finally, in Section 5, we demonstrate the applicability of our proposed new model to a case study taken from the field of bio-manufacturing. Throughout the text, we use lower case Latin characters for indices, uppercase Latin bold letters for sets, uppercase Latin characters for variables, and Greek letters for parameters.

2. Background

In this section, we present the necessary background to be able to follow through the new general state-space model that we propose in this paper. Here, first, we layout the general problem statement, a standard problem representation framework, and briefly describe model classification, and solution methods. For a detailed discussion, the reader is referred to the following review papers [1,3,4,26]. Second, we show a mixed integer linear programming (MILP) based widely adopted scheduling model. Third, we describe the typical state-space formulation adopted in model predictive control (MPC) technology. Finally, we provide a short overview of the state-space based scheduling model pioneered by Subramanian and co-workers [24,27,28].

2.1. Chemical Production Scheduling

2.1.1. General Problem Statement

The general scheduling problem can be stated as follows. Given:

(i)   Production facility data (e.g., unit capacities and connectivity),
(ii)   Production recipes (e.g., processing times and mixing rules),
(iii) Production costs (e.g., material holding costs),
(iv)   Material availability (e.g., raw materials delivery amounts and dates),
(v)   Resource availability (e.g., maintenance schedule and utility levels), and
(vi)   Production targets or orders with due-times;

scheduling seeks to find:

(i)   Number and the associated processing-sizes of the needed tasks,
(ii)   Assignment of these tasks to processing units, and
(iii) Timing (or just the sequence) of these tasks on the assigned units;

so as to meet production targets at minimum cost, or to maximize profit if production beyond the given target is allowed. Apart from minimization of cost, or maximization of profit, the objective can also be minimization of makespan, or minimization of earliness, or any other suitable objective for the considered application. In general, several processing characteristics and constraints could also be present such as sequence dependent changeovers, setup times, storage constraints, time-varying utility costs, etc. [1,3].

2.1.2. Problem Representation

Before a scheduling problem can be solved, we need an abstract framework to represent the different elements of the problem, viz., the production facility, the associated production recipe, etc. The state task network (STN) enables this representation [29]. Under this representation, tasks are carried out on units (equipment), and they transform materials (states) from one to another. Apart from the material to be processed and the equipment to process these materials on, these tasks can also require resources, such as, utilities, manpower etc. Another popular framework, is the resource task network (RTN) [30]. In contrast to STN, in which materials, units, and utilities are treated as different from one another, in RTN, these are treated at par, all termed together as resources. We use the STN representation in this paper, but the general modeling ideas presented are also easily adaptable to the RTN representation.

The STN representation primarily comprises of tasks $i \in \mathbf{I}$, units $j \in \mathbf{J}$, and materials $k \in \mathbf{K}$. The set of tasks producing/consuming material k are denoted by $\mathbf{I}_k^+/\mathbf{I}_k^-$; task i consumes/produces material k equivalent to $\rho_{ik}/\bar{\rho}_{ik}$ mass fraction of its batch-size ($\rho_{ik} < 0$ for consumption and $\bar{\rho}_{ik} > 0$ for production). The subset of tasks that can be carried out on unit j are denoted by $\mathbf{I}_j$; The processing time of task i, when executed on unit j, is denoted by τ_{ij}. On any given unit, only one task can be performed at a time with its batch-size between lower (β_{ij}^{min}) and upper capacities (β_{ij}^{max}); the associated fixed and proportional production costs of carrying out task i on unit j are α_{ij}^F and α_{ij}^P respectively. Feed, intermediate, and product materials are denoted by $k \in \mathbf{K}^F/\mathbf{K}^I/\mathbf{K}^P$; there are possible incoming deliveries (ζ_{kt}) and outgoing orders ($\bar{\zeta}_{kt}$) at certain times for selected materials; the selling price, inventory cost, and backlog cost of material k are γ_k, γ_k^{INV}, and γ_k^{BO}, respectively.

In Figure 1, we see a process network's STN representation comprising of 4 material nodes (circular) labeled M0-M3 and 4 task nodes (rectangular) labeled T1-T4. Arcs connect task nodes with corresponding input/output material nodes. Tasks can be carried out in compatible units and could require utilities. Task-unit mapping and task batch-size capacities (β^{min}/β^{max}) are also shown here. Material prices (γ) are shown adjacent to the material nodes.

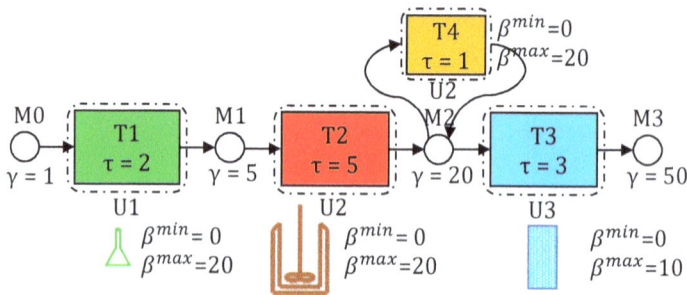

Figure 1. STN representation of a process network. This network, for a bio-manufacturing process, is described in detail in Section 5. It has three steps (tasks) for production of a pharmaceutical ingredient (material M3). T1 is the task of preparing the cell cultures in lab-size beakers. T2 denotes the task of having the cell culture grow and produce the pharmaceutical active ingredient, on a feed of sugars, in a bio-reactor. T3 is a purification task, which is carried out in chromatograph columns. Finally, T4 is a dummy task to model storage of material M2 inside unit U2. $\rho_{ik}/\bar{\rho}_{ik}$ values (not shown in the figure) are either -1 or $+1$ depending on whether the task is consuming the material or producing it.

2.1.3. Model Classification

Scheduling models can be classified on the basis of (i) optimization decisions; (ii) modeling elements; and (iii) modeling of time [4]. Models that employ a time-grid are either continuous time or discrete time models. In discrete time models, the fixed time-grid spacing is denoted by δ. Events can take place only at these grid time-points. Thus, all time-related parameters are rounded in a conservative direction, such that the resulting schedule computed using these new parameter values is feasible even for the original parameter values. Hence, processing times and raw material delivery dates are rounded up, while due dates are rounded down so as to match with an integer multiple of δ.

Even though, having a discrete time-grid introduces the above approximation error, discrete time models have several advantages over continuous time-grid models. For example, accounting for utility consumption, inventory and backlog costs, time varying prices, or time-dependent resource availability introduces non-linearities in continuous time models, but not so in discrete time models [26]. Furthermore, discrete time models are, in general, at least as effective as continuous time models, and in fact are better suited for large scale problems with several additional processing features [31]. In this work, we employ a discrete time-grid for our state-space model.

2.1.4. Solution Methods

To tackle the computational challenge of MILP scheduling models, several solution methods have been proposed: (1) tightening methods based on preprocessing algorithms and valid inequalities [32–38]; (2) reformulations [5,37,39–41]; (3) decomposition methods [42–47]; (4) heuristics [19,48–50]; and (5) hybrid methods [51–55]. Finally, parallel computing has been utilized to obtain faster solutions [56–58].

2.2. Scheduling MILP Model

The discrete time STN MILP scheduling model modified from Shah et al. (1999) [59] comprises of Equations (1)–(6). Time is represented by index $t \in \mathbf{T}$. Binary variable W_{ijt}, when 1, implies task i is starting on unit j at time t. Variable $B_{ijt} \in [\beta_{ij}^{min}, \beta_{ij}^{max}]$ denotes its batch-size. The assignment constraint (Equation (1)) ensures only one task can be executed on a unit at a time.

$$\sum_{i \in \mathbf{I}_j} W_{ijt} + \sum_{i \in \mathbf{I}_j} \sum_{n=1}^{\tau_{ij}-1} W_{ij(t-n)} \leq 1 \quad \forall j, t \tag{1}$$

Equation (2) ensures that the batch-size of a task, if initiated, is within its upper and lower bounds.

$$\beta_{ij}^{min} W_{ijt} \leq B_{ijt} \leq \beta_{ij}^{max} W_{ijt} \quad \forall j, i \in \mathbf{I}_j, t \tag{2}$$

S_{kt}, which is the variable denoting inventory of material k during time-period $(t-1, t]$, is calculated in Equation (3) as a balance of production/consumption and outgoing (V_{kt})/incoming (ζ_{kt}) shipments.

$$S_{k(t+1)} = S_{kt} + \sum_{j} \sum_{i \in \mathbf{I}_j \cap \mathbf{I}_k^+} \bar{\rho}_{ik} B_{ij(t-\tau_{ij})} + \sum_{j} \sum_{i \in \mathbf{I}_j \cap \mathbf{I}_k^-} \rho_{ik} B_{ijt} - V_{kt} + \zeta_{kt} \quad \forall k, t \tag{3}$$

Equation (4) couples the outgoing shipment variable V_{kt} with demand, $\tilde{\zeta}_{kt}$, for material k at time t. Backlog variables, BO_{kt}, denote pending demand during time-period $(t-1, t]$, and are penalized in the cost minimization objective function (Equation (5)).

$$BO_{k(t+1)} = BO_{kt} - V_{kt} + \tilde{\zeta}_{kt} \quad \forall k, t \tag{4}$$

$$z_{cost} = \min \sum_{k} \sum_{t} (\gamma_k^{INV} S_{kt} + \gamma_k^{BO} BO_{kt}) + \sum_{j} \sum_{i \in \mathbf{I}_j} \sum_{t} (\alpha_{ij}^F W_{ijt} + \alpha_{ij}^P B_{ijt}) \tag{5}$$

Finally, the domain of all the variables is restricted via Equation (6):

$$W_{ijt} \in \{0, 1\}; B_{ijt}, V_{kt}, S_{kt}, BO_{kt} \geq 0 \tag{6}$$

2.3. Standard form of State-Space Models

State-space model formulations have been useful, alongside frequency domain models, in process control [60–64]. Now, as optimization based control and economic MPC are becoming the new standard, state-space models have become ubiquitous [65,66]. In the most general form, a state-space based model can be written as $\frac{dx}{dt} = f(x, u, d)$; where x are the states, u are the manipulated inputs, and d are the disturbances. The function $f(\cdot)$ is not theoretically restricted to the class of linear functions, but is typically approximated as linear due to computational tractability considerations. The linear difference equation form for $f(\cdot)$ yields the model as:

$$x(t+1) = Ax(t) + Bu(t) + B_d d(t) \tag{7}$$

where, A, B, and B_d are state-space matrices and t is the index for time. The states x need not be associated with a physically identifiable entity in the plant. Some can have a direct physical meaning, while others can be artificial (e.g., augmented) constructs so as to enable the modeling exercise. The output (measurements y) is related to the states and inputs as $y = h(x, u)$, where $h(\cdot)$ can be non-linear, but is typically linear (e.g., $y(t) = Cx(t) + Du(t)$, where C and D are coefficient matrices). The control optimization model has to follow the plant physical constraints and any other imposed constraints due to operational strategy (e.g., for environmental concerns) or those that enable better closed-loop properties (e.g., economics and stability). These constraints, when linear, can take the general form:

$$E_x x(t) + E_u u(t) + E_d d(t) \leq 0 \tag{8}$$

where, E_x, E_u, and E_d are the coefficient matrices of the states, inputs, and disturbances, respectively. If there are any equality constraints, these can also be represented as two opposite inequality constraints, so as to conform to the general form (Equation (8)). For example, the following constraints are equivalent:

$$(E_x x(t) + E_u u(t) + E_d d(t) = 0) \Leftrightarrow \left(\begin{bmatrix} E_x \\ -E_x \end{bmatrix} x(t) + \begin{bmatrix} E_u \\ -E_u \end{bmatrix} u(t) + \begin{bmatrix} E_d \\ -E_d \end{bmatrix} d(t) \leq 0 \right) \tag{9}$$

Thus, any equality constraints that we propose from here on, can be easily converted to the general inequality form through the use of the above trick. Finally, the objective function takes the form:

$$z_{cost} = \min_{u(0), u(1), \dots, u(N-1)} V_N(x(0), u(0), x(1), u(1), \dots, x(N-1), u(N-1)) \tag{10}$$

where N is the number of discrete time-points in the online optimization horizon.

A wealth of literature focuses on the closed-loop properties of the aforementioned iterative control methods, with novel and most recent results, specifically, in presence of discrete inputs, discussed in Rawlings and Risbeck (2017) [67].

2.4. Scheduling State-Space Model

Motivated by process control approaches, Subramanian et al. (2012) [24] proposed a state-space model (Equations (5) and (11)–(16)) for the chemical production scheduling problem. For brevity, we present the formulation for constant batch-sizes ($\beta_{ij}^{min} = \beta_{ij}^{max} = \beta_{ij}$). There are two distinct features of this model. First, the "complete status" of the plant can be interpreted solely from the variables (states) at that moment in time. This is made possible by *lifting* past actions/inputs (the task start binary variables, W_{ijt}) which have a lagged effect on the "current status" of the plant. Second, observed uncertainties are treated as disturbances, and represented as parameters in the model equations. These two features, together, allow for the model to be kept identical in each online scheduling iteration without any ad-hoc adjustments (due to observation of uncertainty). Thus, the model is in "online ready" form. In addition, due to the use of the state-space formulation, which is popular in process control models, this model also happens to be a very suitable candidate for integration of scheduling and control [68].

To enable lifting of inputs, new task-states (variables) $\bar{W}^n_{ijt}$ are defined. Although this increases the number of variables in the model, it is matched by an equal increase in the number of equations (the lifting equations, Equations (11) and (12)). Thus, no new degrees of freedom are introduced. When the task starts, n is zero ($n = 0$, Equation (11)), and when the task finishes, n equals the processing time of the task ($n = \tau_{ij}$). To express task delay and unit breakdown disturbances, new parameters $\hat{Y}^n_{ijt}$ and $\hat{Z}^n_{ijt}$ are defined, respectively. $\hat{Y}^n_{ijt}$, when 1, denotes a delay of δ h in task i during time-period $[t - \delta, t)$, where δ, as defined in Section 2.1.3, is the granularity of the discrete time-grid. $\hat{Z}^n_{ijt}$, when 1, denotes break-down of unit j while executing task i during time-period $[t - \delta, t)$. For ease of presentation, we assume from here on that $\delta = 1$ h.

$$\bar{W}^0_{ijt} = W_{ijt} \quad \forall j, i \in \mathbf{I}_j, t \tag{11}$$

$$\bar{W}^n_{ij(t+1)} = \bar{W}^{n-1}_{ijt} - \hat{Y}^{n-1}_{ijt} + \hat{Y}^n_{ijt} - \hat{Z}^{n-1}_{ijt} \quad \forall j, i \in \mathbf{I}_j, t, n \in \{1, 2, ..., \tau_{ij}\} \tag{12}$$

In the absence of delays or breakdowns, the lifting equations effectively represent the relation: $\bar{W}^n_{ijt} = W_{ij(t-n)} \quad \forall j, i \in \mathbf{I}_j, n$. The lifted variables are defined only till $n = \tau_{ij}$, because a "look-back" beyond that value of n is not needed. The effect of past inputs, for $n > \tau_{ij}$, is already, indirectly, contained in the inventory and backlog variables S_{kt} and BO_{kt}. The lifted states, $\bar{W}^n_{ijt}$, are augmented to the future states (see Figure 2).

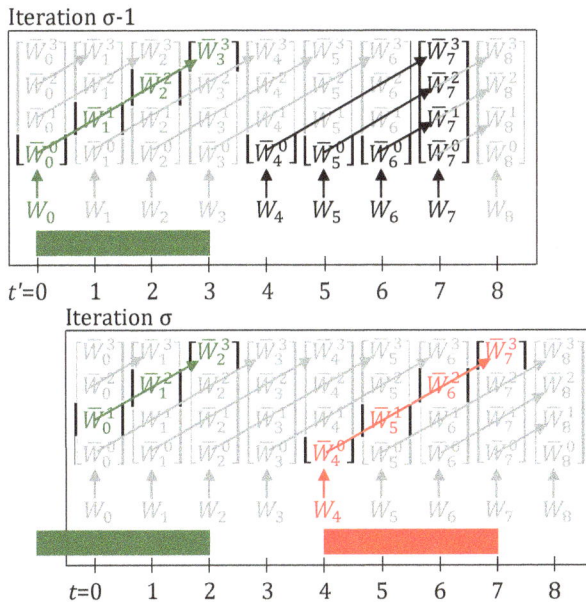

Figure 2. Task-states are shown for two online iterations – numbered $\sigma - 1$ and σ. Each iteration uses its own local time-grid which is reset to start from 0. Here, τ_{ij} for the tasks is assumed to be 3. Lifting of past inputs enables knowing the complete status of the plant by looking at the states (variables) only at that moment in time. In the absence of delays or breakdowns, the lifting equations effectively represent the relation: $\bar{W}^n_{ijt} = W_{ij(t-n)} \quad \forall j, i \in \mathbf{I}_j, n$. Arrows show which variables are equal due to the lifting equations (Equations (11) and (12), with no delays or breakdowns). Variables in green or red have a value of 1, rest have value 0. Information is carried over from one iteration to the next through the update step (Equations (17)–(19)).

In the assignment constraint (Equation (13)), parameters $\hat{Y}_{ijt}^{\tau_{ij}}$ and $\hat{Z}_{ijt}^{\tau_{ij}}$ are included, to ensure that the unit appears to be busy, and no new tasks can be started, when there is a delay or breakdown observed at a time when a task is about to finish. Additionally, for multi-period breakdowns, parameter $\hat{Z}_{\mathrm{IT},jt}^{\tau_{\mathrm{IT},j}}$ is made 1, where IT is a fictitious "idle task", with $\tau_{\mathrm{IT},j} = 1$, that keeps the unit busy through the duration of the multi-period breakdown.

$$\sum_{i \in I_j} W_{ijt} + \sum_{i \in I_j} \sum_{n=1}^{\tau_{ij}-1} \bar{W}_{ijt}^n + \sum_{i \in I_j} (\hat{Y}_{ijt}^{\tau_{ij}} + \hat{Z}_{ijt}^{\tau_{ij}}) \leq 1 \quad \forall j, t \tag{13}$$

In inventory balance (Equation (14)), $\hat{\beta}_{ijkt}^C$ and $\hat{\beta}_{ijkt}^P$ are parameters that denote material handling loss during consumption and production of material k, respectively. When a delay or breakdown is observed at the end of a task, the terms $\hat{Y}_{ijt}^{\tau_{ij}}$ and $\hat{Z}_{ijt}^{\tau_{ij}}$, which are subtracted from $\bar{W}_{ijt}^{\tau_{ij}}$, prevent erroneous multiple counting of the material amount produced by that task.

$$S_{k(t+1)} = S_{kt} + \sum_{j} \sum_{i \in I_j \cap I_k^+} (\bar{\rho}_{ik}\beta_{ij}(\bar{W}_{ijt}^{\tau_{ij}} - \hat{Y}_{ijt}^{\tau_{ij}} - \hat{Z}_{ijt}^{\tau_{ij}}) + \hat{\beta}_{ijkt}^P) \tag{14}$$

$$+ \sum_{j} \sum_{i \in I_j \cap I_k^-} (\rho_{ik}\beta_{ij}W_{ijt} + \hat{\beta}_{ijkt}^C) - V_{kt} + \zeta_{kt} \quad \forall k, t$$

In the backorder balance (Equation (15)), $\hat{\zeta}_{kt}$ denotes demand disturbance.

$$BO_{k(t+1)} = BO_{kt} - V_{kt} + \hat{\zeta}_{kt} \quad \forall k, t \tag{15}$$

Finally, Equation (16) shows the bounds on the variables present in the model.

$$W_{ijt}, \bar{W}_{ijt}^n \in \{0, 1\}; \ S_{kt}, BO_{kt}, V_{kt} \geq 0 \tag{16}$$

Next, we describe the online update step, i.e., how information is carried over from one online iteration to the next. Since the scheduling horizon is advanced by 1 h (the model is kept identical), the state at $t = 0$ (initial condition) for the next iteration is matched with the state at $t' = 1$ of the previous iteration. This is shown in Figure 2, and achieved through the online "update equations" (Equations (17)–(19)), in which σ denotes the iteration number. Variables $_\sigma S_{k(t=0)}$, $_\sigma BO_{k(t=0)}$, and $_\sigma \bar{W}_{ij(t=0)}^n$ for $n \geq 1$ which represent lifted task-states, are assigned fixed values through the update step. But, $_\sigma \bar{W}_{ij(t=0)}^0$, which represents degrees of freedom to start new tasks at $t = 0$, is not fixed. This is identical to how the online updates are performed for the no disturbance case [25]. However, since here we are dealing with the case where disturbances can be present, the disturbance parameters ($\hat{\zeta}_{kt}$, $\hat{Y}_{ijt}^n$, $\hat{Z}_{ijt}^n$, $\hat{\beta}_{ijkt}^P$, and $\hat{\beta}_{ijkt}^C$) are also assigned appropriate values to reflect the observed disturbances. However, these parameters do not participate in the update equations. These influence the prediction of states, for $t \geq 1$, in the online iteration σ.

$$_\sigma \bar{W}_{ij(t=0)}^n = {}_{(\sigma-1)}\bar{W}_{ij(t'=1)}^n \quad \forall j, i \in I_j, n \in \{1, 2, ..., \tau_{ij}\} \tag{17}$$

$$_\sigma S_{k(t=0)} = {}_{(\sigma-1)} S_{k(t'=1)} \quad \forall k \tag{18}$$

$$_\sigma BO_{k(t=0)} = {}_{(\sigma-1)} BO_{k(t'=1)} \quad \forall k \tag{19}$$

Figure 3A,B, show the evolution of task-states when a 2 h delay is observed right after a task starts and just before a task is about to finish, respectively. The 2 h duration of this multi-period delay is known immediately in iteration σ. However, the model formulation also does allow for representing the observation of consecutive, possibly independent, 1 h single-period delays, one at a time in succeeding iterations. These collectively, in hindsight, appear to be a single multi-period delay, but are actually not.

It is quite evident, that n, now in the presence of delays, loses its physical meaning of denoting how much progress has been made on the task. For example, for the task in Figure 3A (iteration σ), due to the 2 h delay, the task-states evolve as $\bar{W}^1_{(t=0)} = \bar{W}^1_{(t=1)} = \bar{W}^1_{(t=2)} = \bar{W}^2_{(t=3)} = \bar{W}^3_{(t=4)} = 1$, instead of the more intuitive $\bar{W}^0_{(t=0)} = \bar{W}^0_{(t=1)} = \bar{W}^1_{(t=2)} = \bar{W}^2_{(t=3)} = \bar{W}^3_{(t=4)} = 1$. Similarly, in Figure 3B (iteration σ), the task-states evolve as $\bar{W}^3_{(t=0)} = \bar{W}^3_{(t=1)} = \bar{W}^3_{(t=2)} = 1$, instead of the more intuitive $\bar{W}^2_{(t=0)} = \bar{W}^2_{(t=1)} = \bar{W}^3_{(t=2)} = 1$. All that can be said now is that when $\bar{W}^0_{ijt} = 1$, the task has just started at time t, and when $\bar{W}^{\tau_{ij}}_{ijt} = 1$ and $\hat{Y}^{\tau_{ij}}_{ijt} = 0$ simultaneously, then the task has finished. As we will show in Section 3.1, we overcome this limitation by introducing a new convention to map observed disturbances to the disturbance parameters, and hence, are able to preserve the physical meaning of n, even when disturbances are present (see Figure 4).

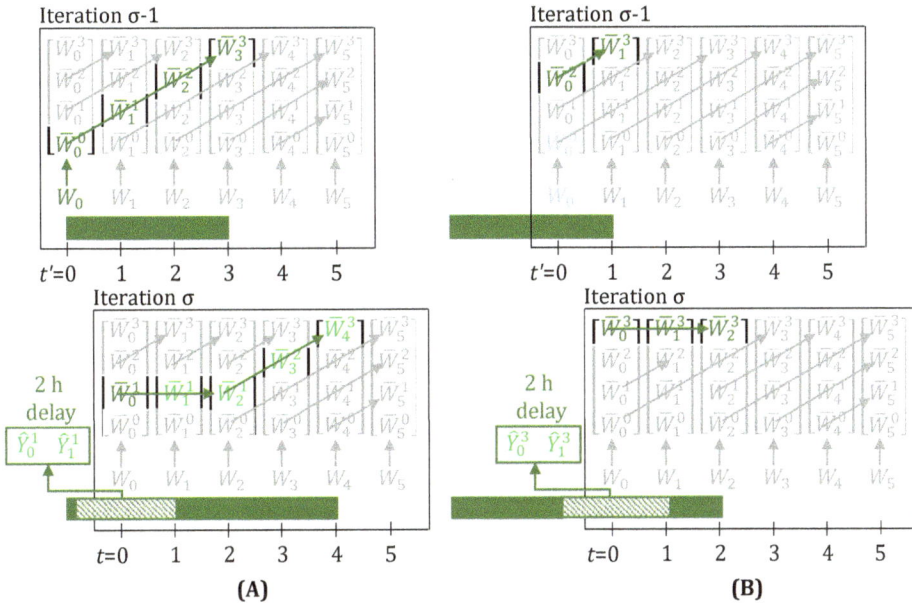

Figure 3. When a 2 h delay is observed, through the lifting equations, $\bar{W}^n_{ijt}$ evolve over the green trajectory, leading to the task correctly finishing 2 h late in iteration σ. Here, τ for the task is 3. Arrows show which variables are enforced as equal by the lifting equations (Equations (11) and (12), with delays present). Variables and parameters in green have a value of 1, rest have value 0. (**A**) The task now finishes at $t = 4$, instead of at $t = 2$. (**B**) The task now finishes at $t = 2$, instead of at $t = 0$. Through Equation (13), the unit is kept busy at $t = 0$ and 1, by the inclusion of the terms $\hat{Y}^{\tau=3}_{(t=0)}$ and $\hat{Y}^{\tau=3}_{(t=1)}$, and hence a new task is prevented from starting at these times. In addition, these terms in Equation (14), prevent the task's produce from erroneously contributing to inventory ($S_{k,(t=1)}$ and $S_{k,(t=2)}$).

Figure 5A,B, show the evolution of task-states when a breakdown just before $t = 0$ is observed and is known to have a 2 h unit downtime (for repairs), right after a task starts and just before a task is about to finish, respectively. Given the observation of breakdown, we would expect intuitively, and unlike what is shown in Figure 5A,B, that none of the task-states $\bar{W}^n_{ijt}$ are active for the green task at $t = 0$. We show how this is achieved through the new model discussed in Section 3.1 (see Figure 6).

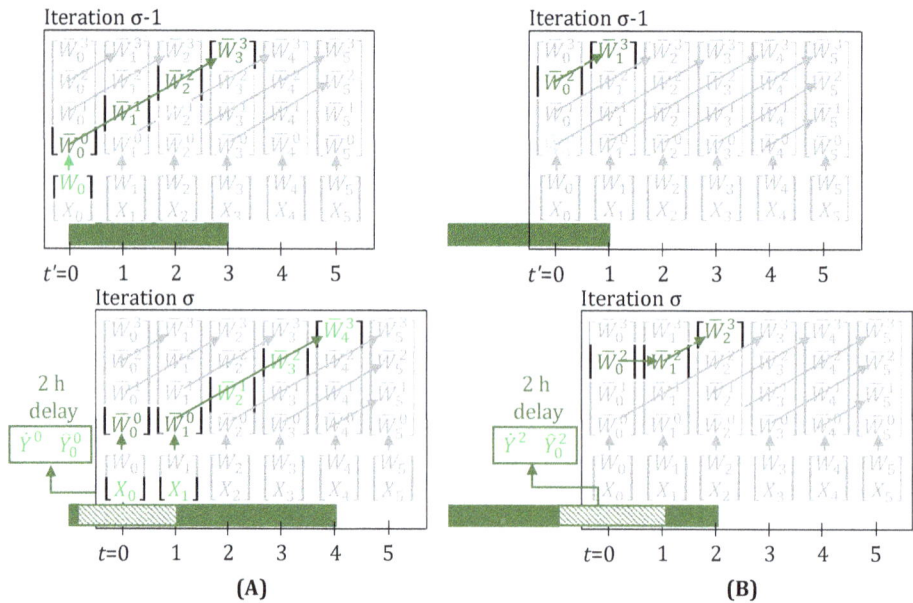

Figure 4. When a 2 h delay is observed, through the lifting equations, $\bar{W}_{ijt}^n$ evolve over the green trajectory, leading to the task correctly finishing 2 h late in iteration σ. Here, τ for the task is 3. Arrows show which variables are enforced as equal by the lifting equations (Equations (22)–(24), with delays present). Variables and parameters in green have a value of 1, rest have value 0. (**A**) The task now finishes at $t = 4$, instead of at $t = 2$. Due to the update step, parameters $\hat{Y}$, $\hat{Y}_0^0$ and variable X_0 are 1, hence, in iteration σ, due to the optimization model, $\bar{W}_0^0$, X_1, and $\bar{W}_1^0$ are also 1. (**B**) The task now finishes at $t = 2$, instead of at $t = 0$. The update step ensures that the true progress, n, of the task is reflected in the task-states at $t = 0$, i.e., $n = 2$.

3. Modeling Generalizations

In Section 3.1, we present a new state-space model formulation that differs, from the state-space model of Subramanian et al. (2012) [24], in the convention that is followed for mapping observed disturbances to the disturbance parameters. Although both models are accurate, this new convention ensures that the task-states, in the presence of disturbances, follow a more intuitive notation. Specifically, the meaning of n as the progress of a task, is maintained. In addition, we define several new parameters to systematically account for disturbances.

In Section 3.2, we show how to handle fractional delays and unit downtimes (due to unit breakdowns). In Section 3.3, we expand the scope of the model to account for variable batch-sizes. Thereafter, in Sections 3.4–3.9 we present generalizations that can be applied to the state-space model, one at a time. Afterwards, in Section 4, we present the final model equations with all generalizations present simultaneously. As we will see in that section, for all the generalizations to work in the presence of each other, a few more modifications are necessary.

3.1. New Basic Formulation

The new state-space model relies on a comprehensive update step of the task-states, in between the online iterations, to promptly reflect the delays and breakdowns in the task-states. The inventory and backorder update stay the same (Equations (18) and (19)) as in the model of Subramanian et al. (2012) [24]. The task-states update is modified from Equation (17) to Equations (20) and (21).

$$\sigma \bar{W}^n_{ij(t=0)} = {}_{(\sigma-1)}\bar{W}^{n-1}_{ij(t'=0)} - \dot{Y}^{n-1}_{ij} + \dot{Y}^n_{ij} - \dot{Z}^{n-1}_{ij} \quad \forall j, i \in \mathbf{I}_j, n \in \{1, 2, ..., \tau_{ij}\} \tag{20}$$

$$\sigma X_{ij(t=0)} = \dot{Y}^0_{ij} \quad \forall j, i \in \mathbf{I}_j \tag{21}$$

The parameters $\dot{Y}^n_{ij}$ which, if 1, represent a 1 h delay in task with progress status n. Note the dot ($\cdot$) instead of the hat ($\wedge$) on the symbols of these parameters. Since these parameters are exclusively for the update step, and do not directly participate in the optimization model, these need not be indexed by time—neither t' (iteration $\sigma - 1$) nor t (iteration σ). Similarly, $\dot{Z}^n_{ij}$ denotes a breakdown of unit j on which task i with progress status n was running. X_{ijt} is a new binary variable, defined for all time-points, which, when 1, captures the information about delays in a task with progress status $n = 0$, i.e., when the task gets delayed right after it starts. The use of this variable, in Equations (22) and (23), will become clear when we discuss the optimization model. We also define a new parameter $\hat{\Lambda}_{jt}$, which, when 1, denotes the unit is unavailable for the time-period $[t, t+1)$. This parameter participates in Equation (25).

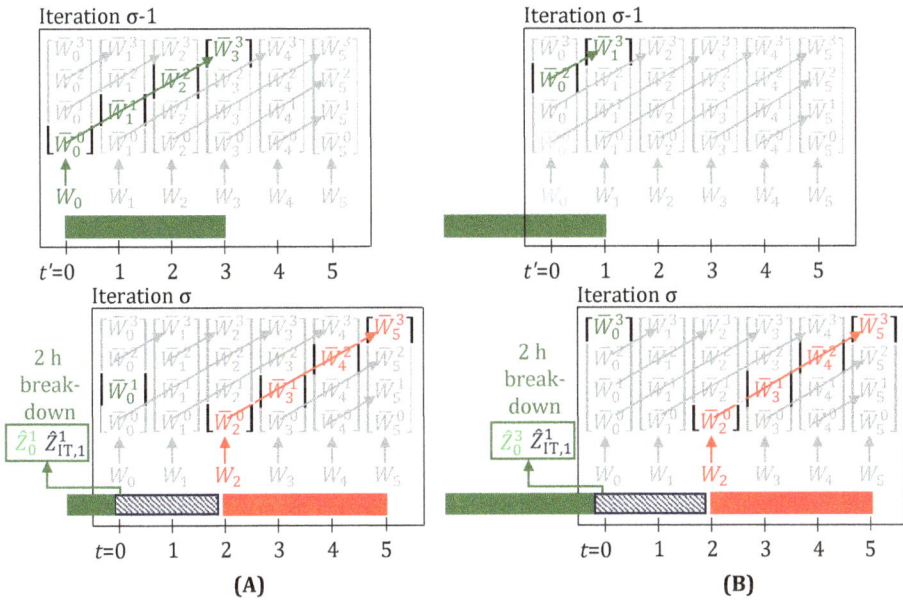

Figure 5. When a breakdown is observed, further evolution of the task-states for the task (the green trajectory), running on the unit that broke, stops. Here, τ for the task is 3 and the unit downtime (blue) is 2 h. Arrows show which variables are enforced as equal by the lifting equations (Equations (11) and (12), with breakdown present). Variables and parameters in green, blue, or red have a value of 1, rest have value 0. The green task is suspended at $t = 0$. A new task (red) can only start at $t = 2$, once the unit downtime is over. (**A**) Through Equation (13), the unit is kept busy at $t = 0$ and 1, due to the terms $\bar{W}^1_{t=0}$ and $\hat{Z}^1_{\mathrm{IT},1}$, respectively. (**B**) Through Equation (13), the unit is kept busy at $t = 0$ and 1, due to the terms $\hat{Z}^{\tau=3}_{t=0}$ and $\hat{Z}^{\tau=1}_{\mathrm{IT},1}$, respectively. Additionally, the term $\hat{Z}^{\tau=3}_{(t=0)}$ in Equation (14), prevents the green task-state ($\bar{W}^{\tau_{ij}=3}_{t=0}$) from erroneously contributing to the inventory.

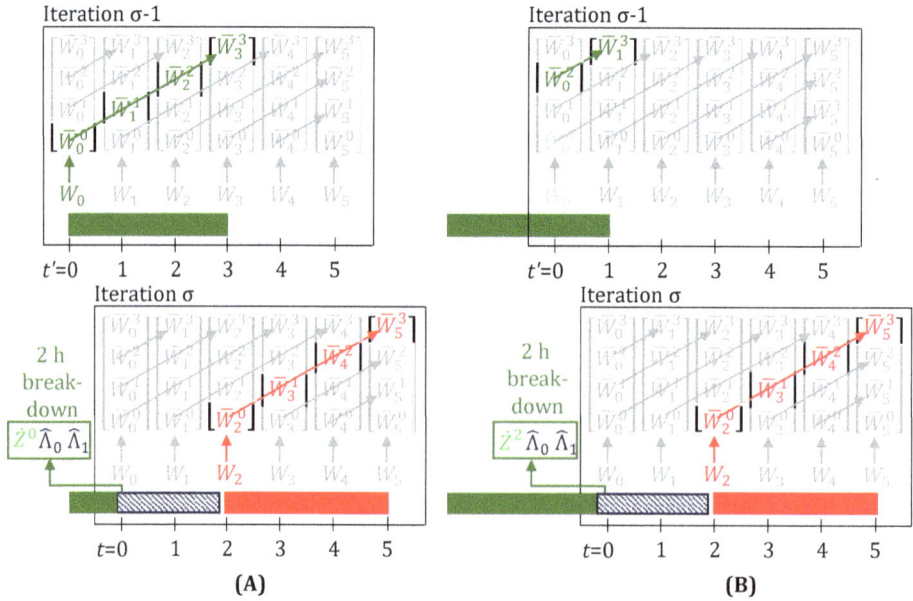

Figure 6. When a breakdown is observed, further evolution of the task-states for the task (the green trajectory), running on the unit that broke, stops. Here, τ for the task is 3 and the unit downtime (blue) is 2 h. Arrows show which variables are enforced as equal by the lifting equations (Equations (22)–(24)). Variables and parameters in green, blue, or red have a value of 1, rest have value 0. The green task is suspended at $t = 0$. A new task (red) can only start at $t = 2$, once the unit downtime is over. Through Equation (25), the unit is kept busy at $t = 0$ and 1, by the inclusion of the terms $\hat{\Lambda}_{t=0}$ and $\hat{\Lambda}_{t=1}$. (**A**) The parameter $\dot{Z}^0$, through Equation (20), prevents the task-state from evolving from $_{(\sigma-1)}\bar{W}_0^0$ to $_\sigma\bar{W}_0^1$. (**B**) The parameter $\dot{Z}^2$, through Equation (20), prevents the task-state from evolving from $_{(\sigma-1)}\bar{W}_0^2$ to $_\sigma\bar{W}_0^3$.

For a multi-period delay of ϕ h, in task i on unit j, in addition to $\dot{Y}_{ij}^n$, parameters $\hat{Y}_{ijt}^n$ are activated for $t = 0, t = 1, ..., t = \phi - 2$. For unit breakdowns with downtime duration of ϕ h, in addition to $\dot{Z}_{ij}^n$, parameters, $\hat{\Lambda}_{jt}$ are activated for $t = 0, t = 1, ..., t = \phi - 1$. Thus, single-period delays do not result in activation of any $\hat{Y}_{ijt}^n$ parameters, but single-period breakdowns require activation of $\hat{\Lambda}_{j(t=0)}$.

Having described the update step, we now describe the optimization model. In this model, the lifting equations consist of Equations (22)–(24).

$$X_{ij(t+1)} = \hat{Y}_{ijt}^0 \quad \forall j, i \in \mathbf{I}_j, t \tag{22}$$

$$\bar{W}_{ijt}^0 = W_{ijt} + X_{ijt} \quad \forall j, i \in \mathbf{I}_j, t \tag{23}$$

$$\bar{W}_{ij(t+1)}^n = \bar{W}_{ijt}^{n-1} - \hat{Y}_{ijt}^{n-1} + \hat{Y}_{ijt}^n \quad \forall j, i \in \mathbf{I}_j, t, n \in \{1, 2, ..., \tau_{ij}\} \tag{24}$$

When there is a ϕ h multi-period delay in a task with progress $n = 0$, the update step assigns $X_{ij(t=0)} = 1$ and $\hat{Y}_{ijt}^0 = 1 \ \forall t \in \{0, 1, ..., \phi - 2\}$. This ensures that $\bar{W}_{ijt}^0$ stays activated for next $(\phi-1)$ h, but with $W_{ijt} = 0$, i.e., the task is not erroneously interpreted as a new task start. If there are no delays, then, through Equations (21) and (22), $X_{ijt} = 0$ and any new task that starts with $W_{ijt} = 1$, through Equation (23), results in $\bar{W}_{ijt}^0 = 1$. Equation (23) is a constraint that we impose on the inputs (W_{ijt}) given the states (X_{ijt} and $\bar{W}_{ijt}^0$), and if needed can be converted to inequality form through use of Equation (9). Variables X_{ijt} are either fixed ($t = 0$) by the update step or are equated to the delay parameters in the optimization model, hence, can be declared as free variables with no explicit bounds.

The assignment constraint (Equation (25)) includes the parameter $\hat{\Lambda}_{jt}$ to account for unit downtime. Additionally, it contains the variable $\bar{W}^{0}_{ijt}$ on the left-hand side, and not variable W_{ijt}, to correctly account for the unit being busy, specifically, when a delay in a task with progress $n = 0$ is observed.

$$\sum_{i \in I_j} \bar{W}^{0}_{ijt} + \sum_{i \in I_j} \sum_{n=1}^{\tau_{ij}-1} \bar{W}^{n}_{ijt} \leq 1 - \hat{\Lambda}_{jt} \quad \forall j, t \tag{25}$$

The inventory balance, Equation (26), in contrast to Equation (14), does not require any corrective delay or breakdown terms. This is because, for any task, the states W_{ijt} (task-start) and $\bar{W}^{\tau_{ij}}_{ijt}$ (task-end), even if delays or breakdowns are observed, are active only at most once.

$$S_{k(t+1)} = S_{kt} + \sum_{j} \sum_{i \in I_j \cap I_k^+} (\bar{\rho}_{ik}\beta_{ij}\bar{W}^{\tau_{ij}}_{ijt} + \hat{\beta}^{P}_{ijkt}) \tag{26}$$
$$+ \sum_{j} \sum_{i \in I_j \cap I_k^-} (\rho_{ik}\beta_{ij}W_{ijt} + \hat{\beta}^{C}_{ijkt}) - V_{kt} + \zeta_{kt} \quad \forall k, t$$

The complete optimization model consists of Equations (5), (15), (16), and (22)–(26). Figures 4 and 6, respectively, show the evolution of task-states when delays or breakdowns are observed.

3.2. Fractional Delays and Unit Downtimes

In Figures 4 and 6, we showed the cases where delays and unit downtime are integer multiples of time-grid spacing δ. Additionally, the unit breakdown was assumed to take place at almost the time-point t, i.e., very close to an integer multiple of δ. However, if δ is not very small, then these assumptions may not be good. Given any fractional delays (π_{delay}), downtimes (π_{down}), or unit breakdown time (π_{break}), we need an appropriate scheme for the (online iterations) update step, to ensure realistic rounding of these to integer values, so as to keep the task-finish and unit-availability times, in sync with the discrete time-grid. A single task can have multiple separate delays, hence, we index the delay time with index r (recurrence), i.e., π^{r}_{delay}. A breakdown, however, can occur only once, at π_{break}, following which, the unit downtime, π_{down}, starts.

For the first delay, a rounded up value is applied in the update steps, i.e., the delay is assumed to be $\lceil \pi^{1}_{delay}/\delta \rceil$. For every additional ψ^{th} delay, the difference, $\phi = \lceil (\sum_{r=1}^{\psi} \pi^{r}_{delay})/\delta \rceil - \lceil (\sum_{r=1}^{\psi-1} \pi^{r}_{delay})/\delta \rceil$, dictates how much additional, integer ϕ, delay is applied in the update steps. Figure 7A shows a numerical example for fractional delays.

When a unit breaks down, the parameter $\dot{Z}^{n}$ is always activated, so as to suspend the running task. The key challenge is to identify, for how many next time-points the unit would be unavailable. This dictates, if, and how many, $\hat{\Lambda}_t$ parameters are activated. This is done as follows. On breakdown, the unit becomes unavailable from π_{break} to $\pi_{break} + \pi_{down}$ (in iteration σ, $\pi_{break} < 0$). Hence, all $\hat{\Lambda}_t$ that span integer multiple of δ, $t \in (\pi_{break}, \pi_{break} + \pi_{down}]$ are activated. This also means, if $(\pi_{break} - \lfloor \pi_{break}/\delta \rfloor \delta + \pi_{down}) < \delta$, none of the $\hat{\Lambda}_t$ are activated, i.e., the unit breaks down and comes back online before the immediate next time-point. This is illustrated in Figure 7B.

3.3. Variable Batch-Sizes

To account for variable batch-sizes, we define variables, B_{ijt} which denotes the batch-size of the task that just starts, $\hat{B}^{n}_{ijt}$ for lifted task batch-size states, and $^{B}X_{ijt}$ to represent batch-size of task that is delayed with progress status $n = 0$. We define parameters $^{B}\dot{Y}_{ij}$ and $^{B}\dot{Z}^{n}_{ij}$ that participate in the update steps and these denote the batch-size of the task delayed and suspended due to unit break-down, respectively. Further, we define parameter $^{B}\hat{Y}^{n}_{ijt}$ for the optimization model.

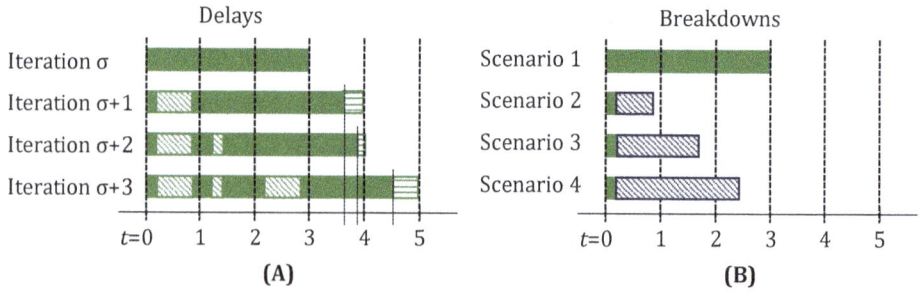

Figure 7. The task has a nominal processing time $\tau = 3$. Time-grid spacing is $\delta = 1$ h. Here, for ease of discussion, the time-grid is global, i.e., it is not reset for each iteration. (**A**) A delay (shown as oblique green pattern) of 0.66 h in time-period $(0, 1)$ is observed. Since, $\lceil 0.66 \rceil = 1$, a delay of 1 h is applied at $t = 1$. This would ensure that the task finish is aligned with $t = 4$, even if the task actually ends at $t = 3.66$. The horizontal green pattern represents the fictitious extra task runtime to align with the discrete time-grid. Next, another delay of 0.2 h is observed in time-period $(1, 2)$. Since, $\lceil 0.66 + 0.2 \rceil - \lceil 0.66 \rceil = 0$, no additional delay is applied in the update steps. This makes sense, because the task finishes at $t = 3.86$ in reality. Since, the previous delay was applied as 1 h, the task is now still thought to finish at $t = 4$, in alignment with the time-grid. Finally, when another delay of 0.66 h is observed in time-period $(2, 3)$. Since, $\lceil 0.66 + 0.2 + 0.66 \rceil - \lceil 0.66 + 0.2 \rceil = 1$, a 1 h delay is applied in the update step. This correctly ensures that the task is now thought to end at $t = 5$, which is the round up of the true end time of $t = 4.52$ h. (**B**) A break-down is observed at $t = 0.2$. Thus, $\dot{Z}^0 = 1$ for the update step between the iterations starting at $t = 0$ and $t = 1$. If the downtime (blue) is 0.66 h, then the unit actually becomes available at $t = 0.86$. Thus, $\hat{\Lambda}_{(t=1)}$ is not activated. This is indeed the case from our mathematical procedure as well, since, $t \in (0.2, 0.86]$ does not include any integer time-point. If $\pi_{down} = 1.5$ h, then $t \in (0.2, 1.7]$ does span $t = 1$, and consequently $\hat{\Lambda}_{(t=1)} = 1$. Finally, if $\pi_{down} = 2.25$ h, then $t \in (0.2, 2.45]$ spans $t = 1$ and $t = 2$, which results in $\hat{\Lambda}_{(t=1)} = 1$ and $\hat{\Lambda}_{(t=2)} = 1$.

The additional update steps, Equations (27) and (28), due to variable batch-sizes, are as follows:

$$_{\sigma}\bar{B}^n_{ij(t=0)} = {}_{(\sigma-1)}\bar{B}^{n-1}_{ij(t'=0)} - {}^{B}\dot{Y}^{n-1}_{ij} + {}^{B}\dot{Y}^{n}_{ij} - {}^{B}\dot{Z}^{n-1}_{ij} \quad \forall j, i \in \mathbf{I}_j, n \in \{1, 2, ..., \tau_{ij}\} \tag{27}$$

$$_{\sigma}^{B}X_{ij(t=0)} = {}^{B}\dot{Y}^{0}_{ij} \quad \forall j, i \in \mathbf{I}_j \tag{28}$$

The optimization model now requires Equations (29)–(31) for lifting the batch-size:

$$^{B}X_{ij(t+1)} = {}^{B}\hat{Y}^{0}_{ijt} \quad \forall j, i \in \mathbf{I}_j, t \tag{29}$$

$$\bar{B}^{0}_{ijt} = B_{ijt} + {}^{B}X_{ijt} \quad \forall j, i \in \mathbf{I}_j, t \tag{30}$$

$$\bar{B}^{n}_{ij(t+1)} = \bar{B}^{n-1}_{ijt} - {}^{B}\hat{Y}^{n-1}_{ijt} + {}^{B}\hat{Y}^{n}_{ijt} \quad \forall j, i \in \mathbf{I}_j, t, n \in \{1, 2, ..., \tau_{ij}\} \tag{31}$$

It might appear in Equation (30) that when $^{B}X_{ijt} > 0$, nothing prevents B_{ijt} from also erroneously taking on a positive value. This was not an issue in Equation (23) because the W_{ijt} and $\bar{W}^0_{ijt}$ variables there were binary. However, Equation (2) ensures that B_{ijt} can only take a non-zero value when $W_{ijt} = 1$. Since, through Equation (23), $W_{ijt} = 0$, whenever $X_{ijt} = 1$, B_{ijt} also takes value 0. The update steps ensure that X_{ijt} and $^{B}X_{ijt}$ can only be non-zero simultaneously.

The inventory balance (Equation (32)) now incorporates the new batch-size variables, B_{ijt} and $\bar{B}^n_{ijt}$, rather than the task-state binary variables (W_{ijt} and $\bar{W}^n_{ijt}$) which was the case in Equation (26).

$$S_{k(t+1)} = S_{kt} + \sum_{j} \sum_{i \in \mathbf{I}_j \cap \mathbf{I}^+_k} (\bar{\rho}_{ik} \bar{B}^{\tau_{ij}}_{ijt} + \hat{\beta}^{P}_{ijkt}) + \sum_{j} \sum_{i \in \mathbf{I}_j \cap \mathbf{I}^-_k} (\rho_{ik} B_{ijt} + \hat{\beta}^{C}_{ijkt}) - V_{kt} + \zeta_{kt} \quad \forall k, t \tag{32}$$

Finally, the variable bounds are as follows:

$$W_{ijt}, \bar{W}^n_{ijt} \in \{0,1\}; B_{ijt}, \bar{B}^n_{ijt}, S_{kt}, BO_{kt}, V_{kt} \geq 0 \tag{33}$$

The update step comprises of Equations (18)–(21), (27) and (28), and the optimization model consists of Equations (2), (5), (15), (22)–(25) and (29)–(33).

Remark: In principle, we can completely avoid defining the new parameters ${}^B\dot{Y}_{ij}$, ${}^B\dot{Z}^n_{ij}$, ${}^B\hat{Y}_{ijt}$, and variable ${}^BX_{ijt}$ by reformulating Equations (27)–(31), so as to only use parameters $\dot{Y}_{ij}$, $\dot{Z}^n_{ij}$, $\hat{Y}_{ijt}$, and variable X_{ijt}. For example, Equation (31) can be reformulated to Equation (34).

$$\bar{B}^n_{ij(t+1)} = \bar{B}^{n-1}_{ijt}(1 - \hat{Y}^{n-1}_{ijt} + \hat{Y}^n_{ijt}) \quad \forall j, i \in \mathbf{I}_j, t, n \in \{1, 2, ..., \tau_{ij}\} \tag{34}$$

Since, Equation (34) entails the multiplication of variables with parameters, by itself, it is an acceptable alternate linear formulation. However, when *task termination* is allowed as a scheduling decision (Section 3.7), this reformulation results in bi-linear terms which are undesirable. Thus, we indeed define the new parameters ${}^B\dot{Y}_{ij}$, ${}^B\dot{Z}^n_{ij}$, ${}^B\hat{Y}_{ijt}$, and variable ${}^BX_{ijt}$, and use Equations (27)–(31) in their native form without the simplifying reformulation discussed in this remark.

3.4. Robust Scheduling: Batch-Sizes

In many applications, it can be prudent to schedule batches bigger than what are needed to just satisfy the nominal demand. This can be, for example, due to the possibility of seeing a demand spike, or to pro-actively compensate for typical material handling losses when a batch finishes. To do so, the parameter $\bar{\rho}_{ik}$ in material inventory balance (Equation (35b)) can be substituted by a scaled down value ($\bar{\rho}^r_{ik}$), where $\bar{\rho}^r_{ik} < \bar{\rho}_{ik}$. This results in bigger batches starting, since the model now under-predicts the yield of materials from any given batch-size. In order to, however, correctly account for the actual inventory resulting from the finishing of a task, the nominal value of $\bar{\rho}_{ik}$ is used at $t = 0$, along with any yield-loss or material handling loss disturbance (Equation (35a)). As it can be seen in Figure 8, which is a simple illustration of this modeling generalization, as the iterations progress, a task-finish-state eventually hits $t = 0$, yielding the large yield proportionate to the true (nominal) value of $\bar{\rho}_{ik}$. Now, if there are any material handling losses ($\hat{\beta}^P_{ijk(t=0)}$), they can be subtracted from the true yield (in Equation (35a)). It is worth noting that, although $\hat{\beta}^P_{ijkt}$ and $\hat{\beta}^C_{ijkt}$ are defined for all time-points, they are possibly active only at $t = 0$, if the corresponding uncertainty is observed. Hence, these parameters can be, in principle, dropped from Equation (35b).

$$S_{k(t+1)} = S_{kt} + \sum_j \sum_{i \in \mathbf{I}_j \cap \mathbf{I}^+_k} (\bar{\rho}_{ik}\bar{B}^{\tau_{ij}}_{ijt} + \hat{\beta}^P_{ijkt}) + \sum_j \sum_{i \in \mathbf{I}_j \cap \mathbf{I}^-_k} (\rho_{ik}B_{ijt} + \hat{\beta}^C_{ijkt}) - V_{kt} + \zeta_{kt} \quad \forall k, t \in \{0\} \tag{35a}$$

$$S_{k(t+1)} = S_{kt} + \sum_j \sum_{i \in \mathbf{I}_j \cap \mathbf{I}^+_k} (\bar{\rho}^r_{ik}\bar{B}^{\tau_{ij}}_{ijt} + \hat{\beta}^P_{ijkt}) + \sum_j \sum_{i \in \mathbf{I}_j \cap \mathbf{I}^-_k} (\rho_{ik}B_{ijt} + \hat{\beta}^C_{ijkt}) - V_{kt} + \zeta_{kt} \quad \forall k, t \in \{1,2,3,...\} \tag{35b}$$

We can write the above two equations, compactly together, as follows:

$$S_{k(t+1)} = S_{kt} + \sum_j \sum_{i \in \mathbf{I}_j \cap \mathbf{I}^+_k} (\theta^{\rho}_{ikt}\bar{B}^{\tau_{ij}}_{ijt} + \hat{\beta}^P_{ijkt}) \tag{36}$$

$$+ \sum_j \sum_{i \in \mathbf{I}_j \cap \mathbf{I}^-_k} (\rho_{ik}B_{ijt} + \hat{\beta}^C_{ijkt}) - V_{kt} + \zeta_{kt} \quad \forall k, t$$

where

$$\theta^{\rho}_{ikt} = \begin{cases} \bar{\rho}_{ik} & \forall k, t = \{0\}; \\ \bar{\rho}^r_{ik} & \forall k, t = \{1,2,3,...\} \end{cases} \tag{37}$$

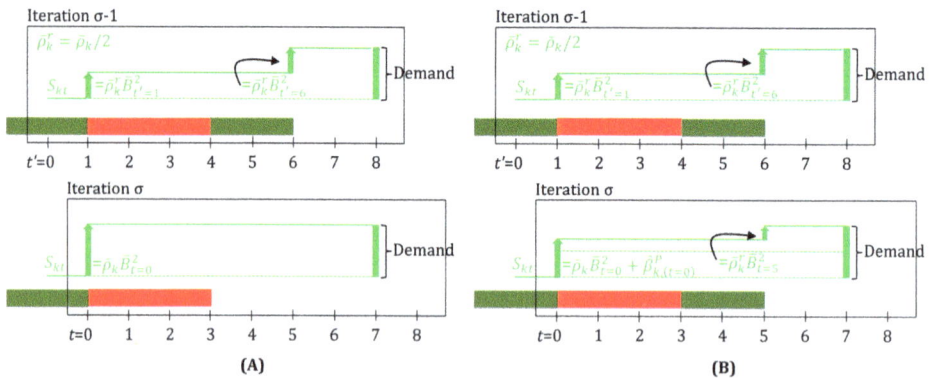

Figure 8. In iteration $\sigma - 1$, the green task ($\tau = 2$) with a "large batch" is finishing at $t' = 1$, but, due to the use of $\bar{\rho}_k^r$, is anticipated to produce only half of what the demand is. Thus, another identical green task is scheduled to start at $t' = 4$, to satisfy the demand. With the use of $\bar{\rho}_k^r$, if the demand could be still be satisfied with a single "large" batch, a second batch wouldn't be scheduled. In iteration σ, the earlier green task yields a large amount of material, in line with its large batch-size due to the true value of $\bar{\rho}_{ik}$ used at $t = 0$. (**A**) Since, here, there was no material handling loss, thus the second green task need not run now, as the demand was satisfied by the first batch itself. (**B**) Although the green task results in a large yield at $t = 0$, a small material handling loss ($\beta_{k(t=0)}^P < 0$) at $t = 0$, requires the second green task to still be scheduled in order to meet the demand, but now with a smaller batch-size. If there are no further material handling losses, there would be a small excess inventory of material produced by the second green task, since its batch-size was decided assuming the yield to be lower ($\bar{\rho}_k^r$) but would in reality by higher ($\bar{\rho}_k$).

3.5. Robust Scheduling: Processing Times

Uncertainty in the processing times is very common in scheduling [69]. A popular approach to proactively manage this uncertainty is to *robustify* the schedule by adding a delay buffer to each task's processing time [70]. Once this robust schedule has been computed, it is advantageous to adjust it online, by taking into account the feedback on actual finish times of the tasks [71]. In discrete-time models, this has been typically done using ad-hoc adjustments in between the online iterations. To the best knowledge of the authors, there is not yet a systematic way to be able to naturally handle this adjustment within an optimization model.

We show here how we can extend the state-space model to produce robust schedules, from a processing time point of view, and yet seamlessly allow for tasks to finish, after they have been running for their nominal processing times plus the delays. We define a new parameter τ_{ij}^r, which denotes the conservative processing time of the tasks ($\tau_{ij}^r > \tau_{ij}$). Thereafter, we modify the lifting (Equation (24) modified to Equations (38a), (38b) and (31) modified to Equations (39a) and (39b)), assignment (Equation (25) modified to Equations (40a) and (40b)), and inventory balance equations (Equation (32) modified to Equations (41a) and (41b)), such that the nominal value of processing times (τ_{ij}) is employed at $t = 0$, and the conservative value (τ_{ij}^r) is employed for $t > 0$. No other model or update equations are modified. An illustration is given in Figure 9.

$$\bar{W}_{ij(t+1)}^n = \bar{W}_{ijt}^{n-1} - \hat{Y}_{ijt}^{n-1} + \hat{Y}_{ijt}^n \quad \forall j, i \in \mathbf{I}_j, n \in \{1,2,...,\tau_{ij}\}, t \in \{0\} \tag{38a}$$

$$\bar{W}_{ij(t+1)}^n = \bar{W}_{ijt}^{n-1} - \hat{Y}_{ijt}^{n-1} + \hat{Y}_{ijt}^n \quad \forall j, i \in \mathbf{I}_j, n \in \{1,2,...,\tau_{ij}^r\}, t \in \{1,2,3,...\} \tag{38b}$$

$$\bar{B}_{ij(t+1)}^n = \bar{B}_{ijt}^{n-1} - {}^B\hat{Y}_{ijt}^{n-1} + {}^B\hat{Y}_{ijt}^n \quad \forall j, i \in \mathbf{I}_j, n \in \{1,2,...,\tau_{ij}\}, t \in \{0\} \tag{39a}$$

$$\bar{B}_{ij(t+1)}^n = \bar{B}_{ijt}^{n-1} - {}^B\hat{Y}_{ijt}^{n-1} + {}^B\hat{Y}_{ijt}^n \quad \forall j, i \in \mathbf{I}_j, n \in \{1,2,...,\tau_{ij}^r\}, t \in \{1,2,3,...\} \tag{39b}$$

$$\sum_{i \in I_j} \bar{W}^0_{ijt} + \sum_{i \in I_j} \sum_{n=0}^{\tau_{ij}-1} \bar{W}^n_{ijt} \leq 1 - \hat{\Lambda}_{jt} \quad \forall j, t \in \{0\} \tag{40a}$$

$$\sum_{i \in I_j} \bar{W}^0_{ijt} + \sum_{i \in I_j} \sum_{n=0}^{\tau^r_{ij}-1} \bar{W}^n_{ijt} \leq 1 - \hat{\Lambda}_{jt} \quad \forall j, t \in \{1, 2, 3, \dots\} \tag{40b}$$

$$S_{k(t+1)} = S_{kt} + \sum_{j} \sum_{i \in I_j \cap I_k^+} (\bar{\rho}_{ik} \bar{B}^{\tau_{ij}}_{ijt} + \hat{\beta}^P_{ijkt}) \tag{41a}$$

$$+ \sum_{j} \sum_{i \in I_j \cap I_k^-} (\rho_{ik} B_{ijt} + \hat{\beta}^C_{ijkt}) - V_{kt} + \zeta_{kt} \quad \forall k, t \in \{0\}$$

$$S_{k(t+1)} = S_{kt} + \sum_{j} \sum_{i \in I_j \cap I_k^+} (\bar{\rho}_{ik} \bar{B}^{\tau^r_{ij}}_{ijt} + \hat{\beta}^P_{ijkt}) \tag{41b}$$

$$+ \sum_{j} \sum_{i \in I_j \cap I_k^-} (\rho_{ik} B_{ijt} + \hat{\beta}^C_{ijkt}) - V_{kt} + \zeta_{kt} \quad \forall k, t \in \{1, 2, 3, \dots\}$$

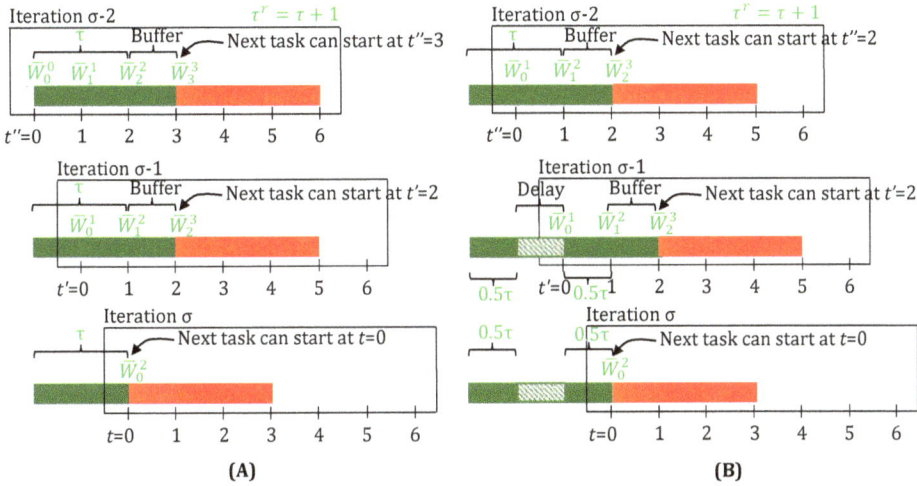

Figure 9. For the green task ($\tau = 2$), a delay buffer of 1 h is chosen, i.e., $\tau^r_{ij} = \tau_{ij} + 1$. (**A**) In iteration $\sigma - 1$, the model predicts that the green task will finish at $t' = 2$. In the next iteration σ, when there is no delay in the green task, having run for 2 h, it correctly finishes at $t = 0$. No ad-hoc model changes are required in between the iterations to make this possible. Consequently, to take advantage of this nominal finish time of the green task, the red task, now starts 1 h earlier than what it was scheduled for in the previous iteration. (**B**) The green task, with conservative processing time, is scheduled to finish at $t'' = 2$ in iteration $\sigma - 2$. In iteration $\sigma - 1$, a delay of 1 h is observed. The processing time delay buffer is maintained, and now the task is anticipated to finish at $t' = 2$. In the next iteration (σ), the task correctly finishes at $t = 0$, accounting for the 1 h delay. The red task can now start on time at $t = 0$, or equivalently $t'' = 2$. Thus, having a delay buffer was useful, since it predicted a realistic start time for the red task in iteration $\sigma - 2$ itself.

The lifting equations, Equations (38b) and (39b), contain variables $W^n_{ij(t=1)}$, $B^n_{ij(t=1)}$ $\forall j, i \in I_j, n \in \{\tau_{ij} + 1, \tau_{ij} + 2, \dots, \tau^r_{ij}\}$, which are not coupled back to any variables at $t = 0$. These variables must be fixed to zero, otherwise, the optimization can assign these variables a spurious value so as to erroneously generate inventory (in Equation (41b) through variables $B_{ijt}^{\tau_{ij}^r} \ \forall j, i \in \mathbf{I}_j, t \in \{1, 2, ..., \tau_{ij}^r - \tau_{ij}\}$).

We can write the above equations, compactly together, as follows:

$$\bar{W}_{ij(t+1)}^n = \bar{W}_{ijt}^{n-1} - \hat{Y}_{ijt}^{n-1} + \hat{Y}_{ijt}^n \ \ \forall j, i \in \mathbf{I}_j, t, n \in \{1, 2, ..., \theta_{ijt}^\tau\} \tag{42}$$

$$\bar{B}_{ij(t+1)}^n = \bar{B}_{ijt}^{n-1} - {}^B\hat{Y}_{ijt}^{n-1} + {}^B\hat{Y}_{ijt}^n \ \ \forall j, i \in \mathbf{I}_j, t, n \in \{1, 2, ..., \theta_{ijt}^\tau\} \tag{43}$$

$$\sum_{i \in \mathbf{I}_j} \bar{W}_{ijt}^0 + \sum_{i \in \mathbf{I}_j} \sum_{n=0}^{\theta_{ijt}^\tau - 1} \bar{W}_{ijt}^n \le 1 - \hat{\Lambda}_{jt} \ \ \forall j, t \tag{44}$$

$$S_{k(t+1)} = S_{kt} + \sum_j \sum_{i \in \mathbf{I}_j \cap \mathbf{I}_k^+} (\bar{\rho}_{ik} \bar{B}_{ijt}^{\theta_{ijt}^\tau} + \hat{\beta}_{ijkt}^P) \tag{45}$$

$$+ \sum_j \sum_{i \in \mathbf{I}_j \cap \mathbf{I}_k^-} (\rho_{ik} B_{ijt} + \hat{\beta}_{ijkt}^C) - V_{kt} + \zeta_{kt} \ \ \forall k, t$$

where

$$\theta_{ijt}^\tau = \begin{cases} \tau_{ij} & \forall j, i \in \mathbf{I}_j, t = \{0\}; \\ \tau_{ij}^r & \forall j, i \in \mathbf{I}_j, t = \{1, 2, 3, ...\} \end{cases} \tag{46}$$

Finally, please note that in this approach, as can be seen in Figure 9B (iteration $\sigma - 1$), an a priori fixed buffer time ($\tau_{ij}^r - \tau_{ij}$) is added to the task duration, irrespective of whether delays have already been observed (during the execution of the current task). It can be argued that the buffer was meant to absorb the actual delays, and hence, should be cut back for tasks that actually get delayed. This is a fair critique. However, owing to feedback, the true task-finish is accounted for (see Figure 9A iteration σ), and there is no wasted equipment time due to the unused buffer, which otherwise results in idle time in static robust scheduling approaches.

3.6. Feedback on Yield Estimates

The material handling loss ($\hat{\beta}_{ijkt}^C / \hat{\beta}_{ijkt}^P$) disturbance parameters can be used to represent yield losses as well, but these parameters are assigned a value only at $t = 0$, i.e., when a task actually ends and a material handling loss is observed. In many applications, the actual yield of a task can, in fact, be estimated during the task's execution, and does not always come as a surprise when the batch finishes. To incorporate the information about anticipated yield loss in future, these parameters can be assigned values for $t > 0$. However, this information has to be then carried over from one iteration to the next, with corresponding decrement in the t index of these parameters to reflect the shifting time-grid, till the task finishes. In addition, if the task is delayed, these parameter values also have to be delayed. This requires cumbersome mechanisms to handle this information in between the online iterations.

Adapting the state-space model to lift the yield-loss information forward provides a much more natural way to handle this feedback. Thus, we define a new free variable L_{ijt}^n, which is analogous to the batch-size variable $\bar{B}_{ijt}^n$, but, as we can see in Equation (47), when the task finishes, instead of adding to, it subtracts from the inventory.

$$S_{k(t+1)} = S_{kt} + \sum_j \sum_{i \in \mathbf{I}_j \cap \mathbf{I}_k^+} (\bar{\rho}_{ik} (\bar{B}_{ijt}^{\tau_{ij}} - L_{ijt}^{\tau_{ij}}) + \hat{\beta}_{ijkt}^P) \tag{47}$$

$$+ \sum_j \sum_{i \in \mathbf{I}_j \cap \mathbf{I}_k^-} (\rho_{ik} B_{ijt} + \hat{\beta}_{ijkt}^C) - V_{kt} + \zeta_{kt} \ \ \forall k, t$$

Thus, in a way, this new variable can be thought to be the negative counterpart of the batch-size variable. The yield loss can also be delayed by use of the parameters ${}^L\hat{Y}_{ij}^n$ and ${}^L\hat{Y}_{ijt}^n$ so as to stay in sync with the task-finish time, or nullified in case of a unit breakdown through the parameter $^L\dot{Z}^n_{ij}$. This is achieved using the update steps (Equations (48) and (49)) and the model lifting equations (Equations (50)–(52)).

$$\sigma \bar{L}^n_{ij(t=0)} = {}_{(\sigma-1)}\bar{L}^{n-1}_{ij(t'=0)} + \dot{\lambda}^{n-1}_{ij} - {}^L\dot{Y}^{n-1}_{ij} + {}^L\dot{Y}^n_{ij} - {}^L\dot{Z}^{n-1}_{ij} \quad \forall j, i \in \mathbf{I}_j, n \in \{1, 2, ..., \tau_{ij}\} \tag{48}$$

$$\frac{L}{\sigma}X_{ij(t=0)} = {}^L\hat{Y}^0_{ij} \quad \forall j, i \in \mathbf{I}_j \tag{49}$$

$\dot{\lambda}^n_{ij}$ is a parameter the denotes the "additional" yield loss observed (anticipated) in that iteration (σ). When delays and yield losses are observed simultaneously, then the parameters $^L\dot{Y}^n_{ij}$ and $^L\hat{Y}^n_{ij}$ take the value ${}_{(\sigma-1)}\bar{L}^{n-1}_{ij(t'=0)} + \dot{\lambda}^{n-1}_{ij}$, i.e., the total yield loss up to and including in iteration σ.

$$^LX_{ij(t+1)} = {}^L\hat{Y}^0_{ijt} \quad \forall j, i \in \mathbf{I}_j, t \tag{50}$$

$$\bar{L}^0_{ijt} = {}^LX_{ijt} \quad \forall j, i \in \mathbf{I}_j, t \tag{51}$$

$$\bar{L}^n_{ij(t+1)} = \bar{L}^{n-1}_{ijt} - {}^L\hat{Y}^{n-1}_{ijt} + {}^L\hat{Y}^n_{ijt} \quad \forall j, i \in \mathbf{I}_j, t, n \in \{1, 2, ..., \tau_{ij}\} \tag{52}$$

Please note that there are no L_{ijt} variables in Equation (51). If these variables were to be present, they would have to be fixed to zero. Otherwise the optimization itself can spuriously assign non-zero values to these variables, such that the intended true batch-size is $B_{ijt} - L_{ijt}$.

3.7. Task Termination

In the update step (Equations (20), (21), (27) and (28)) we made an implicit assumption that past decisions (task-states with $n > 0$ at $t = 0$) are fixed. In general, more decisions from the previous iteration can be considered fixed or the deviation from them penalized in the current iteration. This has been suggested in the literature to reduce schedule nervousness [23,72,73].

However, to the best knowledge of the authors, no model to date considers canceling/terminating tasks already underway, as an optimization decision. This is surprising, given that whenever is needed, this would be a natural decision for a human scheduler. For example, to prioritize processing for a new rush order, an unfinished process unrelated to this rush order may need to be terminated. Other routine possibilities for task termination are excessive delays or yield losses, as is commonly the case in bio-manufacturing [74,75]. This decision to terminate should be an outcome of the online optimization so as to best react to the observed disturbances or new information. We define task *termination* as a "willful" decision to discontinue a task. This is in contrast with task *suspension*, which is a "forced" discontinuation of a task as a result of a unit breakdown or loss of utility support. Since, preemption is not customary in chemical processes, we assume a total loss of output of a task that is terminated. This is in agreement with how the output of task suspensions is treated.

This termination of tasks can be achieved by "softening" the initial conditions (task-states at $t = 0$) in an online iteration. We introduce a new binary variable, T^n_{ij}, which, when 1, denotes termination of task i, on unit j, which has run-index n. Since, all variables values for iteration $\sigma - 1$ are now a parameter for iteration σ, we can write the following linear equations to achieve this softening:

$$\sigma\bar{W}^n_{ij(t=0)} = \left({}_{(\sigma-1)}\bar{W}^{n-1}_{ij(t'=0)} - \dot{Y}^{n-1}_{ij}\right)(1 - T^{n-1}_{ij}) \tag{53}$$
$$+ \dot{Y}^n_{ij}(1 - T^n_{ij}) - \dot{Z}^{n-1}_{ij} \quad \forall j, i \in \mathbf{I}_j, n \in \{1, 2, ..., \tau_{ij}\}$$

$$\sigma X_{ij(t=0)} = \dot{Y}^0_{ij}(1 - T^0_{ij}) \quad \forall j, i \in \mathbf{I}_j \tag{54}$$

$$\sigma\bar{B}^n_{ij(t=0)} = \left({}_{(\sigma-1)}\bar{B}^{n-1}_{ij(t'=0)} - {}^B\dot{Y}^{n-1}_{ij}\right)(1 - T^{n-1}_{ij}) \tag{55}$$
$$+ {}^B\dot{Y}^n_{ij}(1 - T^n_{ij}) - {}^B\dot{Z}^{n-1}_{ij} \quad \forall j, i \in \mathbf{I}_j, n \in \{1, 2, ..., \tau_{ij}\}$$

$$\frac{B}{\sigma}X_{ij(t=0)} = {}^B\dot{Y}^0_{ij}(1 - T^0_{ij}) \quad \forall j, i \in \mathbf{I}_j \tag{56}$$

Hence, the update equations (Equations (20), (21), (27) and (28) modified to (53)–(56)) now become part of the model. The update equations for inventory (Equation (18)) and backlog (Equation (19)) stay unmodified, and are not softened.

If we had no disturbances, just softening the initial task-states would have sufficed. However, when we have delays, we have to also ensure that we appropriately nullify the effect of delay parameters, which have been already assigned a value at the start of an iteration (optimization). Thus, wherever the delay parameters $\hat{Y}_{ijt}^n$ appear, we multiply these with $(1 - T_{ij}^n)$. Since the coefficients of the $(1 - T_{ij}^n)$ terms are the delay parameters, when there is no delay, these terms are also, consequently, absent. Parameters $\grave{Y}_{ij}^{\tau_{ij}}$, $^B\grave{Y}_{ij}^{\tau_{ij}}$, $\hat{Y}_{ijt}^{\tau_{ij}}$, and $^B\hat{Y}_{ijt}^{\tau_{ij}}$ are always zero, hence, variables $T_{ij}^{\tau_{ij}}$ do not participate in the model. A unit breakdown implicitly disrupts a task, hence, in such a situation, the question of termination does not arise. Overall, the lifting equations are modified to Equations (57)–(60) with Equations (23) and (30) remaining unchanged.

$$X_{ij(t+1)} = \hat{Y}_{ijt}^0(1 - T_{ij}^0) \quad \forall j, i \in \mathbf{I}_j, t \tag{57}$$

$$\bar{W}_{ij(t+1)}^n = \bar{W}_{ijt}^{n-1} - \hat{Y}_{ijt}^{n-1}(1 - T_{ij}^{n-1}) + \hat{Y}_{ijt}^n(1 - T_{ij}^n) \quad \forall j, i \in \mathbf{I}_j, t, n \in \{1, 2, ..., \tau_{ij}\} \tag{58}$$

$$^BX_{ij(t+1)} = {}^B\hat{Y}_{ijt}^0(1 - T_{ij}^0) \quad \forall j, i \in \mathbf{I}_j, t \tag{59}$$

$$\bar{B}_{ij(t+1)}^n = \bar{B}_{ijt}^{n-1} - {}^B\hat{Y}_{ijt}^{n-1}(1 - T_{ij}^{n-1}) + {}^B\hat{Y}_{ijt}^n(1 - T_{ij}^n) \quad \forall j, i \in \mathbf{I}_j, t, n \in \{1, 2, ..., \tau_{ij}\} \tag{60}$$

We do not index variable T_{ij}^n with time (t), because it serves no purpose to terminate a task in future ($t > 0$). If a task is already under execution and has to be terminated in the future for some reason, it is always better to terminate it right away ($t = 0$).

In addition to including a cost associated with task termination, $\sum_j \sum_{i \in \mathbf{I}_j} \sum_{n=0}^{\tau_{ij}-1} \alpha_{ij}^T T_{ij}^n$, in the objective, we can also enforce a pre-specified unit downtime (τ_j^T) following every task termination, by including a summation term in the assignment constraints for that many time-points:

$$\sum_{i \in \mathbf{I}_j} \sum_{n=0}^{\tau_{ij}-1} \bar{W}_{ijt}^n + \sum_{i \in \mathbf{I}_j} \sum_{n=0}^{\tau_{ij}-1} T_{ij}^n \leq 1 - \hat{\Lambda}_{jt} \quad \forall j, t \in \{0, 1, 2, ..., \tau_j^T - 1\} \tag{61a}$$

$$\sum_{i \in \mathbf{I}_j} \sum_{n=0}^{\tau_{ij}-1} \bar{W}_{ijt}^n \leq 1 - \hat{\Lambda}_{jt} \quad \forall j, t \in \{\tau_j^T, \tau_j^T + 1, ...\} \tag{61b}$$

We can write the above two equations, compactly together, as follows:

$$\sum_{i \in \mathbf{I}_j} \sum_{n=0}^{\tau_{ij}-1} \bar{W}_{ijt}^n + \sum_{i \in \mathbf{I}_j} \sum_{n=0}^{\tau_{ij}-1} \theta_{ijt}^T T_{ij}^n \leq 1 - \hat{\Lambda}_{jt} \quad \forall j, t \tag{62}$$

where

$$\theta_{ijt}^T = \begin{cases} 1 & \forall t = \{0, 1, 2, ..., \tau_{ij}^T - 1\}; \\ 0 & \forall t = \{\tau_{ij}^T, \tau_{ij}^T + 1, \tau_{ij}^T + 2, ...\} \end{cases} \tag{63}$$

An added advantage of the compact form, is that the unit downtime can now be a function of the task that was terminated, i.e., τ_{ij}^T is indexed by i as well, and not just j.

To systematically account for unit downtime resulting from task-termination in a previous iteration, i.e., if $_{(\sigma-1)}T_{ij}^n = 1 \quad \forall j, i \in \mathbf{I}_j, n \in \{1, 2, 3, ..., \tau_{ij} - 1\}$, the parameter $\hat{\Lambda}_{jt}$ has to be activated for $t \in \{0, 1, 2, ..., \tau_j^T - 2\}$ in iteration σ. Thereafter, in each subsequent iterations, the downtime is decremented by 1, and the parameter $\hat{\Lambda}_{jt}$ appropriately activated for the corresponding time-points. Alternatively, we can define a new binary variable and lift it, to keep the unit deactivated for the remaining downtime in subsequent iterations. The new variable, which would be subtracted on the right hand side of Equations (61a) and (61b), in lieu of $\hat{\Lambda}_{jt}$, can be thought of as the unit unavailability variable.

3.8. Post-Production Storage in Unit

Kondili et al. (1993) [29] proposed a formulation for "hold" tasks, which are dummy tasks that can be used to model storage of materials in a processing unit, while waiting to be unloaded. This is especially important for production facilities that follow a no intermediate storage (NIS) policy for certain materials. In the network shown in Figure 1, task T4 is a hold task. The purpose of this task is to keep material M3 residing in unit U3. So as to ensure that the hold task can only store material in a unit that it was originally produced in, we write the state-space version of the original constraint proposed by Kondili et al. (1993) [29] in Equation (64).

$$B_{i',jt} \leq \tilde{B}_{i',jt}^{\tau_{i',j}} + \sum_{i \in I_j} \tilde{B}_{ijt}^{\tau_{ij}} \quad \forall i' \in \mathbf{I}_{\text{HOLD}}, j \in J_{i'}, t \tag{64}$$

where $\mathbf{I}_{\text{HOLD}}$ is the set of hold tasks. When multiple materials are produced in a unit, here we assume that the corresponding hold task emulates the simultaneous storage of all these materials in the unit. The $\tilde{\rho}_{\text{HOLD},k}$ mass-coefficient then dictates in what proportion are the materials released from the unit. If only a certain individual material is held, and the others unloaded, the term $\sum_{i \in I_j} \tilde{\rho}_{ik} \tilde{B}_{ijt}^{\tau_{ij}}$ is substituted with $\sum_{i \in I_k^+ \cap I_j} \tilde{\rho}_{ik} \tilde{B}_{ijt}^{\tau_{ij}}$, and the constraint is written only for that material k which is held. Further, for a perishable material, $\tilde{\rho}_{\text{HOLD},k} < \rho_{\text{HOLD},k}$; that is, a fraction of the material is lost (perishes or deactivates) when held.

3.9. Unit Capacity Degradation and Maintenance

In many processes, such as polymerization reactions, or purification processes, it is not uncommon for the unit capacity to "degrade" after a task has been processed on that unit [76–81]. This can be, for example, due to residue formation (e.g., scaling) or impurity accumulation (e.g., membrane pore blockages). Some degradation is gradual and predictable, while some degradation may occur suddenly and unexpectedly.

To model unit degradation, we define a new non-negative variable, C_{ijt}, which denotes the capacity of the unit j, to perform task i that starts at time t. For an un-degraded unit, C_{ijt} is initialized to the value β_{ij}^{max}. This variable value is passed over from one iteration to another, through the update equation:

$$_{\sigma}C_{ij(t=0)} = {}_{(\sigma-1)}C_{ij(t'=1)} + \dot{\mu}_{ij} \quad \forall j \in J_{\text{MT}}, i \in I_j \tag{65}$$

A new disturbance parameter, $\dot{\mu}_{ij}$, which when negative, represents extent of sudden (unexpected) partial loss in unit capacity. Conversely, a positive value represents renewal of unit capacity. This positive value could be, for example, a result of installing a new repaired unit in place of an old unit that broke down.

Through Equation (66), we define a balance on the unit capacity. $\rho_{ii'j}^C$ is a parameter that denotes the gradual degradation in capacity of unit j to perform task i, due to execution of task i' on that unit. $\rho_{ii'j}^C$ is either negative or zero.

$$C_{ij(t+1)} = C_{ijt} + \sum_{i' \in I_j} \sum_{n=1}^{\tau_{i'j}} \frac{\rho_{ii'j}^C}{\tau_{i'j}} B_{i'jt}^n + \tilde{M}_{ijt}^{\text{TMT},j} \quad \forall j \in J_{\text{MT}}, i \in \{I_j \setminus I_{\text{MT}}\}, t \tag{66}$$

The degraded unit can be typically restored to its full capacity through a maintenance task (e.g., cleaning), which we denote with the abbreviation MT. We define $\mathbf{I}_{\text{MT}}$ as a set of maintenance tasks, and J_{MT} as a set of units which can degrade, and consequently, need a corresponding maintenance task. Further, we add the maintenance task (MT) to the set of tasks I_j that can be performed on unit j. The value of $\bar{M}_{ijt}^{\tau_{MT,j}}$, the variable which we define below, dictates the restored capacity due to completion of a maintenance task.

Like any conventional task, the maintenance task has a start binary ($W_{MT,jt}$) associated with it, which is appropriately lifted. The assignment equation (Equation (25)) ensures that the maintenance task can only run when the unit is not running any other task. Since the maintenance task does not consume or produce materials, the conventional batch-size of this maintenance task, $B_{MT,jt}$ is fixed to zero. Instead, we define a new type of batch-size, $\bar{M}_{ijt}^{n}$, specific to the purpose of this task, which we term as the maintenance-size. Since we assume that only one kind of a maintenance task exists for every unit, the index i in this maintenance-size variable is not MT. This variable denotes, how much capacity to perform task i is restored, when maintenance is performed on unit j. This variable is also lifted, similar to the batch-size variable, and can be delayed or suspended due to breakdowns. Similarly, the parameters $^{M}\hat{Y}_{ij}^{n}$, $^{M}\hat{Y}_{ijt}^{n}$, and $^{M}\hat{Z}_{ij}^{n}$ denote the maintenance-sizes of the maintenance task corresponding to capacity restored to perform task i. The update equations associated with this variable are:

$$_{\sigma}\bar{M}_{ij(t=0)}^{n} = {}_{(\sigma-1)}\bar{M}_{ij(t'=0)}^{n-1} - {}^{M}\hat{Y}_{ij}^{n-1} + {}^{M}\hat{Y}_{ij}^{n} - {}^{M}\hat{Z}_{ij}^{n-1} \quad \forall j \in J_{MT}, i \in \{I_j \setminus I_{MT}\}, n \in \{1, 2, ..., \tau_{MT,j}\} \tag{67}$$

$$_{\sigma}^{M}X_{ij(t=0)} = {}^{M}\hat{Y}_{ij}^{0} \quad \forall j \in J_{MT}, i \in \{I_j \setminus I_{MT}\} \tag{68}$$

The model equations, for lifting, are:

$$^{M}X_{ij(t+1)} = {}^{M}\hat{Y}_{ijt}^{0} \quad \forall j \in J_{MT}, i \in \{I_j \setminus I_{MT}\}, t \tag{69}$$

$$\bar{M}_{ijt}^{0} = M_{ijt} + {}^{M}X_{ijt} \quad \forall j \in J_{MT}, i \in \{I_j \setminus I_{MT}\}, t \tag{70}$$

$$\bar{M}_{ij(t+1)}^{n} = \bar{M}_{ijt}^{n-1} - {}^{M}\hat{Y}_{ijt}^{n-1} + {}^{M}\hat{Y}_{ijt}^{n} \quad \forall j \in J_{MT}, i \in \{I_j \setminus I_{MT}\}, t, n \in \{1, 2, ..., \tau_{MT,j}\} \tag{71}$$

The batch-size of new tasks is upper bounded by the unit capacity variable (Equation (72)). This ensures that only smaller batches can now be processed if $C_{ijt} < \beta_{ij}^{max}$. If a task just finishes, the restored or degraded capacity due to that is also accounted for while upper bounding the task-batchsize B_{ijt}.

$$B_{ijt} \le C_{ijt} + \sum_{i' \in I_j} \frac{\rho_{i'j}^{C}}{\tau_{i'j}} B_{i'jt}^{\tau_{i'j}} + \bar{M}_{ijt}^{\tau_{MT,j}} \quad \forall j \in J_{MT}, i \in \{I_j \setminus I_{MT}\}, t \tag{72}$$

We define the maintenance task with fixed processing time ($\tau_{MT,j}$), and assume that whenever performed, the unit is restored to its full capacity (i.e., $C_{ijt} = \beta_{ij}^{max}$). This requires the maintenance-size, M_{ijt}, to be the difference between the deteriorated unit capacity (C_{ijt}), before the maintenance starts, and the upper capacity limit (β_{ij}^{max}), to which the unit has to be restored to. This is achieved using Equation (73):

$$\beta_{ij}^{max}W_{MT,jt} - C_{ijt} \le M_{ijt} \le \beta_{ij}^{max}W_{MT,jt} \quad \forall j \in J_{MT}, i \in \{I_j \setminus I_{MT}\}, t \tag{73}$$

Finally, we specify the lower and upper bounds for variable C_{ijt}:

$$0 \le C_{ijt} \le \beta_{ij}^{max} \quad \forall j \in J_{MT}, i \in \{I_j \setminus I_{MT}\}, t \tag{74}$$

and define a new parameter α_{ij}^{M}, which denotes the proportional cost of maintenance of a unit. The cost term, in the objective, for a maintenance task is $\sum_{j \in J_{MT}} \sum_{t} (\alpha_{MT,j}^{F} W_{MT,jt} + \sum_{i \in \{I_j \setminus I_{MT}\}} \alpha_{ij}^{M} M_{ijt})$.

4. Integrated Model

In this section, we present the complete model with all generalizations present simultaneously. For brevity, we write index σ in only those equations, where $\sigma - 1$ is also present. Everywhere else, all variables are those of the current iteration σ.

The update equations are Equations (18), (19), and (65). The softened update equations, which are part of the model, are Equations (53)–(56) and (75)–(78).

$$_\sigma \bar{L}^n_{ij(t=0)} = \left(_{(\sigma-1)} \bar{L}^{n-1}_{ij(t'=0)} + \lambda^{n-1}_{ij} - {}^L\dot{Y}^{n-1}_{ij}\right)(1 - T^{n-1}_{ij}) \tag{75}$$
$$+ {}^L\dot{Y}^n_{ij}(1 - T^n_{ij}) - {}^L\dot{Z}^{n-1}_{ij} \quad \forall j, i \in \mathbf{I}_j, n \in \{1, 2, ..., \tau_{ij}\}$$

$$_\sigma^L X_{ij(t=0)} = {}^L\dot{Y}^0_{ij}(1 - T^0_{ij}) \quad \forall j, i \in \mathbf{I}_j \tag{76}$$

$$_\sigma \bar{M}^n_{ij(t=0)} = {}_{(\sigma-1)}\bar{M}^{n-1}_{ij(t'=0)} - {}^M\dot{Y}^{n-1}_{ij}(1 - T^{n-1}_{\text{MT},j}) \tag{77}$$
$$+ {}^M\dot{Y}^n_{ij}(1 - T^n_{\text{MT},j}) - {}^M\dot{Z}^{n-1}_{ij} \quad \forall j \in \mathbf{J}_{\text{MT}}, i \in \{\mathbf{I}_j \setminus \mathbf{I}_{\text{MT}}\}, n \in \{1, 2, ..., \tau_{\text{MT},j}\}$$

$$_\sigma^M X_{ij(t=0)} = {}^M\dot{Y}^0_{ij}(1 - T^0_{\text{MT},j}) \quad \forall j \in \mathbf{J}_{\text{MT}}, i \in \{\mathbf{I}_j \setminus \mathbf{I}_{\text{MT}}\} \tag{78}$$

The lifting equations are Equations (23), (30), (51), (57), (59), (70), and (79)–(84).

$$\bar{W}^n_{ij(t+1)} = \bar{W}^{n-1}_{ijt} - \hat{Y}^{n-1}_{ijt}(1 - T^{n-1}_{ij}) + \hat{Y}^n_{ijt}(1 - T^n_{ij}) \quad \forall j, i \in \mathbf{I}_j, t, n \in \{1, 2, ..., \theta^\tau_{ijt}\} \tag{79}$$

$$\bar{B}^n_{ij(t+1)} = \bar{B}^{n-1}_{ijt} - {}^B\hat{Y}^{n-1}_{ijt}(1 - T^{n-1}_{ij}) + {}^B\hat{Y}^n_{ijt}(1 - T^n_{ij}) \quad \forall j, i \in \mathbf{I}_j, t, n \in \{1, 2, ..., \theta^\tau_{ijt}\} \tag{80}$$

$$^L X_{ij(t+1)} = {}^L\hat{Y}^0_{ijt}(1 - T^0_{ij}) \quad \forall j, i \in \mathbf{I}_j, t \tag{81}$$

$$\bar{L}^n_{ij(t+1)} = \bar{L}^{n-1}_{ijt} - {}^L\hat{Y}^{n-1}_{ijt}(1 - T^{n-1}_{ij}) + {}^L\hat{Y}^n_{ijt}(1 - T^n_{ij}) \quad \forall j, i \in \mathbf{I}_j, t, n \in \{1, 2, ..., \theta^\tau_{ijt}\} \tag{82}$$

$$^M X_{ij(t+1)} = {}^M\hat{Y}^0_{ijt}(1 - T^0_{\text{MT},j}) \quad \forall j \in \mathbf{J}_{\text{MT}}, i \in \mathbf{I}_j, t \tag{83}$$

$$\bar{M}^n_{ij(t+1)} = \bar{M}^{n-1}_{ijt} - {}^M\hat{Y}^{n-1}_{ijt}(1 - T^{n-1}_{\text{MT},j}) \tag{84}$$
$$+ {}^M\hat{Y}^n_{ijt}(1 - T^n_{\text{MT},j}) \quad \forall j \in \mathbf{J}_{\text{MT}}, i \in \{\mathbf{I}_j \setminus \mathbf{I}_{\text{MT}}\}, t, n \in \{1, 2, ..., \theta^\tau_{\text{MT},jt}\}$$

where please note the use of parameter θ^τ_{ijt} in Equations (79), (80), (82), and (84).

The assignment constraint is:

$$\sum_{i \in \mathbf{I}_j} \sum_{n=0}^{\theta^\tau_{ijt}-1} W^n_{ijt} + \sum_{i \in \mathbf{I}_j} \sum_{n=0}^{\theta^\tau_{ijt}-1} \theta^\tau_{ijt} T^n_{ij} \leq 1 - \hat{\Lambda}_{jt} \quad \forall j, t \tag{85}$$

The backlog, inventory, and unit capacity balance are Equations (15), (86), and (87), respectively.

$$S_{k(t+1)} = S_{kt} + \sum_j \sum_{i \in \mathbf{I}_j \cap \mathbf{I}^+_k} \left(\theta^\rho_{ikt}(\bar{B}^{\theta^\tau_{ijt}}_{ijt} - \bar{L}^{\theta^\tau_{ijt}}_{ijt}) + \hat{B}^P_{ijkt}\right) \tag{86}$$
$$+ \sum_j \sum_{i \in \mathbf{I}_j \cap \mathbf{I}^-_k} (\rho_{ik}B_{ijt} + \hat{B}^C_{ijkt}) - V_{kt} + \zeta_{kt} \quad \forall k, t$$

$$C_{ij(t+1)} = C_{ijt} + \sum_{i' \in \mathbf{I}_j} \sum_{n=1}^{\theta^\tau_{i'jt}} \frac{\rho^C_{ii'j}}{\theta^\tau_{i'jt}} \bar{B}^n_{i'jt} + \bar{M}^{\theta^\tau_{\text{MT},jt}}_{ijt} \quad \forall j \in \mathbf{J}_{\text{MT}}, i \in \{\mathbf{I}_j \setminus \mathbf{I}_{\text{MT}}\}, t \tag{87}$$

Batch-size, post-production storage, and maintenance-size constraints are Equations (2), (88), and (73) and (89).

$$B_{i',jt} \leq \bar{B}^{\theta^\tau_{i',jt}}_{i',jt} + \sum_{i \in \mathbf{I}_j} \bar{B}^{\theta^\tau_{ijt}}_{ijt} \quad \forall i' \in \mathbf{I}_{\text{HOLD}}, j \in \mathbf{J}_{i'}, t \tag{88}$$

$$B_{ijt} \leq C_{ijt} + \sum_{i' \in \mathbf{I}_j} \frac{\rho^C_{ii'j}}{\theta^\tau_{i'j}} \bar{B}^{\theta^\tau_{i'j}}_{i'jt} + \bar{M}^{\theta^\tau_{\text{MT},j}}_{ijt} \quad \forall j \in \mathbf{J}_{\text{MT}}, i \in \{\mathbf{I}_j \setminus \mathbf{I}_{\text{MT}}\}, t \tag{89}$$

Bounds on the variables are enforced by Equations (74) and (90)–(92).

$$W_{ijt}, \bar{W}_{ijt}^n, T_{ij}^n \in \{0,1\}; \; B_{ijt}, M_{ijt}, \bar{B}_{ijt}^n, S_{kt}, BO_{kt}, V_{kt} \geq 0; \tag{90}$$

$$\bar{W}_{ij(t=1)}^n, \bar{B}_{ij(t=1)}^n, L_{ij(t=1)}^n = 0 \; \forall n = \{\tau_{ij} + 1, \tau_{ij} + 2, ..., \tau_{ij}^r\} \tag{91}$$

$$\bar{M}_{ij(t=1)}^n = 0 \; \forall n = \{\tau_{MT,j} + 1, \tau_{MT,j} + 2, ..., \tau_{MT,j}^r\} \tag{92}$$

Variables X_{ijt}, $^BX_{ijt}$, $^LX_{ijt}$, $^MX_{ijt}$ are free variables, however, through the update and model equations, they are always equated to a parameter value, hence, are not degrees of freedom. The decision variables, or in other words—the inputs from a systems perspective [24], are W_{ijt}, B_{ijt}, T_{ij}^n, M_{ijt}, and V_{kt}.

The objective is:

$$\min \; \sum_k \sum_t (\gamma_k^{INV} S_{kt} + \gamma_k^{BO} BO_{kt}) + \sum_j \sum_{i \in I_j} \sum_t (\alpha_{ij}^F W_{ijt} + \alpha_{ij}^P B_{ijt}) \tag{93}$$

$$+ \sum_j \sum_{i \in I_j} \sum_{n=0}^{\tau_{ij}-1} \alpha_{ij}^T T_{ij}^n + \sum_{j \in J_{MT}} \sum_{i \in \{I_j \setminus I_{MT}\}} \sum_t \alpha_{ij}^M M_{ijt}$$

5. Case Study

Bio-manufacturing is a type of manufacturing in which molecules of interest, such as, metabolites, drugs, enzymes, etc., are produced through the use of biological systems such as living micro-organisms [82]. Several commercial sectors, rely on these, for example, pharmaceuticals, food and beverage processing, agriculture, waste treatment, etc. Furthermore, there is an increasing thrust towards finding biological routes for production of bulk chemicals [83]. The use of live systems in bio-manufacturing, however, introduces several operational challenges. These include batch-to-batch variability, parallel growth of both, the desired product as well as undesired toxic byproducts in the same batch, and possible random shocks that can lead to complete failure of a batch [75]. This makes it an interesting area for application of scheduling methods. In this section, we present an example, motivated from bio-manufacturing, to demonstrate all modeling generalizations discussed in Section 3, using the integrated model equations outlined in Section 4.

In general, bio-manufacturing processes can be divided into an upstream bio-reaction (e.g., fermentation) stage and a downstream purification stage [74]. The upstream stage typically consists of two steps: cell culture preparation in the lab and the bio-reaction. These two steps are task T1 and T2, respectively, in the network in Figure 1. The downstream purification stage typically consists of three steps: centrifugation, chromatography, and filtration. Among these three steps, chromatography takes the longest and the chromatograph columns are prone to unpredictable failures. Hence, we assume that chromatography, being the dominant step, is representative of the complete purification stage. This is task T3 in the network in Figure 1. Overall, we choose this simplified system (network) for an easier illustration of the capabilities of our general state-space model.

In our example, as features, we allow for possible delays in the cell culture preparation (T1), and small yield losses in the bio-reaction (T2), including possible substantial yield losses due to sudden cell death. Thus, we carry out robust scheduling, using a conservative processing time (τ^r) for task T1 and a conservative mass-conversion coefficient ($\bar{\rho}^r$), against small yield losses, for task T2. Further, in the downstream stage, we assume that the chromatograph column (U3) loses capacity with usage, part of which is predictable. Executing a maintenance task can restore capacity on the chromatographs. To enforce the no-intermediate storage (NIS) policy for material M2, the inventory variable $S_{M2,t}$ is fixed to zero for all time-points. Raw material, M0, is assumed to be available in an unrestricted supply, as needed. Selected instance parameter values, other than the ones already shown in Figure 1, are outlined in Table 1.

Table 1. Parameter values for the case study.

$\gamma_k^{INV} = 0.15\gamma_k$	$\gamma_k^{BO} = 1.5\gamma_k$	$\alpha_{ij}^F = 1$	$\alpha_{ij}^P = 0$	$\alpha_{ij}^T = 2$	$\alpha_{ij}^M = 0$
$\tau_{U3}^T = 2$	$\tau_{T1,U1}^r = 3$	$\tau_{MT,U3} = 2$	$\bar{\rho}_{M2,T2}^r = 0.9$	$\rho_{T3,T3,U3}^C = -0.2$	$\delta = 1$

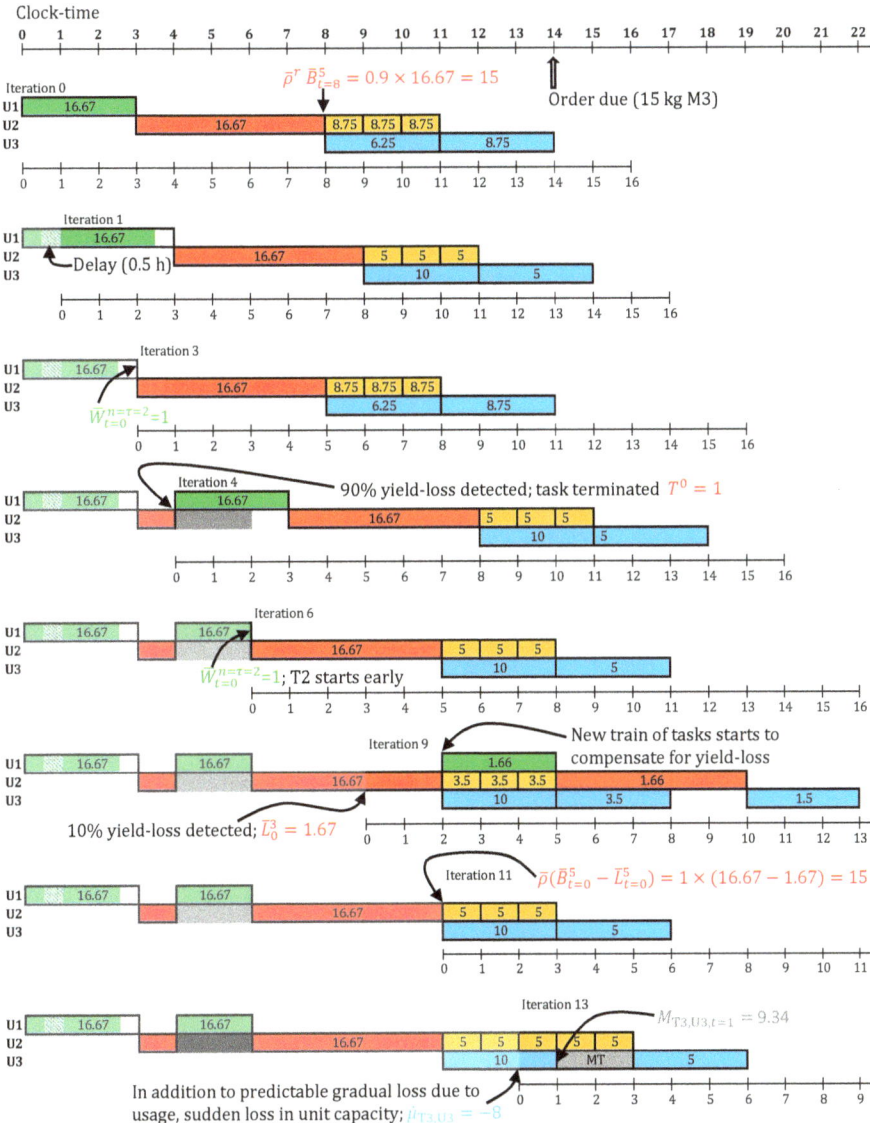

Figure 10. Selected online scheduling iterations for the case study. The color of the tasks corresponds to their color in Figure 1. Executed schedule, with respect to each iteration, is shown in lighter (fainted) colors. The clock-time is the global time. Each iteration has its own local discrete time-grid.

An order for 15 kg of M3 is due at the 14th hour of the day. To meet this order, online scheduling is carried out, with a horizon of 16 h, and re-optimization every 1 h, starting at the 0th hour. All optimizations are solved to optimality using default solver options in CPLEX 12.6.1 (IBM Corporation, North Castle, NY, USA) via GAMS 24.4.3 (GAMS Development Corporation, Fairfax, VA, USA), installed on an Intel Xeon (E5520, 2.27 GHz, 8 core processor) machine (Intel Corporation, Santa Clara, CA, USA), with 16 GB of RAM and Linux CentOS 7 operating system (Red Hat Inc., Raleigh, NC, USA). The schedules obtained in selected online scheduling iteration are shown in Figure 10. For the remaining online iterations, the predicted schedule is identical to the respective previous iteration (but with time-grid shift).

The nominal makespan, without any disturbances or robustification, to meet this order is 13 h. However, in iteration 0 (see Figure 10), T1 is started at $t = 0$, instead of $t = 1$, since a conservative processing time, $\tau^r = 3$ h is in use. Further, the batch-sizes for T1 and T2 are 16.67 kg, since a conservative yield parameter ($\bar{\rho}^r = 0.9$) is in use for T2. This predicts production of 15 kg of M2, which through the two T3 batches, of 6.25 kg and 8.75 kg, makes 15 kg of M3.

In the next iteration, a fractional delay of 0.5 h is observed. Due to the use of a discrete time-grid with granularity $\delta = 1$ h, and being the first delay for this task, this is rounded up to 1 h. Since, now the order is predicted to be late, the batch-sizes of T3 are revised so as to meet as much of the order as possible on time ($\beta^{max} = 10$). Going forward, no more delays are observed in T1, hence, it finishes at $t = 0$ in iteration 3. This is because the nominal τ is in use at $t = 0$ in Equations (85) and (86). Consequently, the downstream tasks are all scheduled earlier now, matching up with the initial predicted schedule in iteration 0. Thus, the conservative processing time was useful, in making T1 start earlier at 0th hour.

In iteration 4, due to sudden cell death, 90% yield loss in T2 is observed (anticipated at task finish). Hence, T2 is terminated. T1 is restarted (with conservative processing time). Since, no delays are observed through the execution of this new task T1, in iteration 6, it finishes after the nominal processing time of 2 h. Thus the start times of the downstream tasks are pulled forward by 1 h.

In iteration 9, 10% yield loss is observed (anticipated at task finish). Since, this is not substantial, T2 is not terminated. Instead, a new train of tasks is scheduled to start at $t = 2$ to compensate for the lost yield. In iteration 11, when T2 finishes, it results in nominal yield ($\bar{\rho}$) minus the 10% anticipated yield loss. Hence, 15 kg of M2 is produced. Thus, the new train of tasks previously scheduled, but not yet started, in iteration 9 are canceled.

In iteration 11, in addition to the gradual decline in chromatograph capacity due to usage ($\rho^C_{T3,T3,U3} < 0$), which does not affect starting the next 5 kg T3 task, a sudden loss in capacity is observed ($\dot{\mu}_{T3,U3} = -8$). Hence, a maintenance task (MT) is scheduled with maintenance-size $M_{T3,T3,t=1} = 9.34$. Once this maintenance is over, the pending task T3 can start. No further disturbances are observed. Consequently, the order is fully met at the 19th hour.

If task termination, conservative processing time, and conservative yield were not used, the order would have been fully met only at the 27th hour (Gantt chart not shown). Hence, using the general state-space model enabled a richer set of decision making, resulting in an overall better schedule.

6. Conclusions

We developed a general state-space model, particularly motivated by an online scheduling perspective, that allows modeling (1) task-delays and unit breakdowns with a new, more intuitive convention over that of Subramanian et al. (2012) [24]; (2) fractional delays and unit downtimes, when using discrete-time grid; (3) variable batch-sizes; (4) robust scheduling through the use of conservative yield estimates and processing times; (5) feedback on task-yield estimates before the task finishes; (6) task termination during its execution; (7) post-production storage of material in unit; and (8) unit capacity degradation and maintenance. Further, we propose a new scheme for updating the state of the process, as well as an overall formulation to enforce constraints (through parameter/variable modifications), based on feedback information, on future decisions. We demonstrate the effectiveness of this model on a case study from the field of bio-manufacturing. Through this new state-space model, we have enabled a natural way to handle routinely encountered processing features and disturbance information in online scheduling. The general features that we address are found in several industrial sectors, namely, pharmaceuticals, fine chemicals, pulp and paper, agriculture, steel production, oil and gas, food processing, bio-manufacturing, etc. The proposed model, therefore, greatly extends and enables the possible application of mathematical programming based online scheduling solutions to diverse application settings. Finally, it is important to note, that although here we presented the model using STN based representation, these generalizations can also be adapted to RTN based representation.

Acknowledgments: The authors acknowledge support from the National Science Foundation under grants CMMI-1334933 and CBET-1264096, as well as the Petroleum Research Fund under grant 53313-ND9. Further, the authors thank Ananth Krishnamurthy for fruitful discussions on bio-manufacturing.

Author Contributions: D.G. conceived the model and prepared the manuscript under the supervision of C.T.M.

Conflicts of Interest: The authors declare no conflict of interest.

Abbreviations

The following abbreviations are used in this manuscript

MILP	mixed integer linear program
MPC	model predictive control
RTN	resource task network
STN	state task network
HOLD	hold (storage) task
MT	maintenance (cleaning) task

Nomenclature

Indices/sets

$i \in \mathbf{I}$	tasks
$j \in \mathbf{J}$	units (equipment)
$k \in \mathbf{K}$	materials
$t \in \mathbf{T}$	time-points/periods
$\mathbf{I} \supseteq \mathbf{I}_j$	tasks that can be carried out in unit j
$\mathbf{I} \supseteq \mathbf{I}_k^+$	tasks producing material k
$\mathbf{I} \supseteq \mathbf{I}_k^-$	tasks consuming material k
$\mathbf{I} \supseteq \mathbf{I}_{\mathrm{HOLD}}$	hold (storage) tasks
$\mathbf{I} \supseteq \mathbf{I}_{\mathrm{MT}}$	maintenance tasks
$\mathbf{J} \supseteq \mathbf{J}_i$	units suitable for carrying out task i
$\mathbf{J} \supseteq \mathbf{J}_{\mathrm{MT}}$	units which can degrade, and consequently, need a corresponding maintenance task
$\mathbf{K} \supseteq \mathbf{K}^F$	feed (raw) materials
$\mathbf{K} \supseteq \mathbf{K}^I$	intermediates
$\mathbf{K} \supseteq \mathbf{K}^P$	final products

Parameters

α_{ij}^F	fixed cost of running task i on unit j
α_{ij}^P	proportional cost of running task i on unit j
α_{ij}^T	cost of terminating task i on unit j
α_{ij}^M	proportional cost of maintenance task i on unit j
β_{ij}	fixed batch-size of task i executed on unit j
$\beta_{ij}^{min} / \beta_{ij}^{max}$	min/max capacity on batch-size of task i executed on unit j
$\hat{\beta}_{ijkt}^P / \hat{\beta}_{ijkt}^C$	material unloading/loading loss during production/consumption of material k
γ_k	selling price of material k
γ_k^{INV}	inventory cost of material k
γ_k^{BO}	backlog cost of material k
δ	discretization of time-grid; length of time-periods
ζ_{kt}	incoming shipment of material k at time t

$\dot{Z}_{ij}^n$	disturbance parameter denoting unit breakdown
$^B\dot{Z}_{ij}^n$	batch-size of task suspended due to unit breakdown
$^L\dot{Z}_{ij}^n$	yield-loss size of task suspended due to unit breakdown
$^M\dot{Z}_{ij}^n$	maintenance-size of maintenance task suspended due to unit breakdown
θ_{ijt}^τ	dummy parameter as defined in Equation (46)
θ_{ikt}^ρ	dummy parameter as defined in Equation (37)
θ_{ijt}^T	dummy parameter as defined in Equation (63)
λ_{ij}^n	yield loss in task i running on unit j, with run status n
$\hat{\Lambda}_{jt}$	binary parameter, when 1, denotes unit j unavailable during time $[t, t+1)$
$\dot{\mu}_{ij}$	when negative, represents extent of sudden partial loss in unit capacity
ξ_{kt}	demand for material k at time t
$\hat{\xi}_{kt}$	demand disturbance for material k at time t
π_{break}	actual (fractional) time at which a unit breaks down
π_{down}	fractional downtime in a unit
π_{delay}	fractional delay in a task
π_{delay}^r	r^{th} delay in a task
ρ_{ik}	mass-conversion coefficient (material consumption)
$\bar{\rho}_{ik}$	mass-conversion coefficient (material production)
$\bar{\rho}_{ik}^r$	conservative mass-conversion coefficient for production ($\bar{\rho}_{ik}^r < \bar{\rho}_{ik}$)
$\rho_{ii'j}^C$	deterioration in unit capacity to perform task i, due to performing task i' on that unit j.
σ	online iteration number
τ_{ij}	processing time of task i on unit j
τ_{ij}^r	conservative processing time of task i ($\tau_{ij}^r < \tau_{ij}$) on unit j
τ_j^T	task independent unit j downtime after terminating a task
τ_{ij}^T	task dependent unit j downtime after terminating task i
$\hat{Y}_{ij}^n, \hat{Y}_{ijt}^n$	single-/multi-period disturbance parameters denoting delay
$^B\hat{Y}_{ij}^n, {}^B\hat{Y}_{ijt}^n$	single-/multi-period disturbance parameters denoting batch-size of a delayed task
$^L\hat{Y}_{ij}^n, {}^L\hat{Y}_{ijt}^n$	single-/multi-period disturbance parameters denoting yield-loss size of a delayed task
$^M\hat{Y}_{ij}^n, {}^M\hat{Y}_{ijt}^n$	single-/multi-period disturbance parameters denoting maintenance-size of delayed maintenance task
ϕ	duration of delay or breakdown, in multiples of δ
ψ	recurrence count of delay for a task

Variables

B_{ijt}	batch-size of task i on unit j
$\bar{B}_{ijt}^n$	lifted batch-size
BO_{kt}	backlog level of material k during period $(t-1, t]$
C_{ijt}	capacity of unit j to perform task i during period $(t-1, t]$
$\bar{L}_{ijt}^n$	lifted yield-loss variables
M_{ijt}	maintenance-size of the maintenance task
$\bar{M}_{ijt}^n$	lifted maintenance-size
S_{kt}	inventory level of material k during period $(t-1, t]$
T_{ij}^n	binary variable, when 1, denotes termination of task i, with run-status n, on unit j
V_{kt}	outgoing shipment to meet demand for material k at time t
W_{ijt}	binary variable, when 1, denotes task i starts on unit j at time-point t
$\bar{W}_{ijt}^n$	lifted task-start variables
X_{ijt}	when 1, captures the information about delays in a task with progress status $n = 0$
$^BX_{ijt}$	the batch-size of delayed task with progress status $n = 0$
$^LX_{ijt}$	yield-loss of delayed task with progress status $n = 0$
$^MX_{ijt}$	maintenance-size of delayed maintenance task with progress status $n = 0$

Reference

1. Harjunkoski, I.; Maravelias, C.T.; Bongers, P.; Castro, P.M.; Engell, S.; Grossmann, I.E.; Hooker, J.; Méndez, C.A.; Sand, G.; Wassick, J.M. Scope for industrial applications of production scheduling models and solution methods. *Comput. Chem. Eng.* **2014**, *62*, 161–193.
2. Kelly, J.D.; Mann, J. Crude oil blend scheduling optimization: An application with multimillion dollar benefits. *Hydrocarb. Process.* **2003**, *82*, 47–54.
3. Méndez, C.A.; Cerdá, J.; Grossmann, I.E.; Harjunkoski, I.; Fahl, M. State-of-the-art review of optimization methods for short-term scheduling of batch processes. *Comput. Chem. Eng.* **2006**, *30*, 913–946.
4. Maravelias, C.T. General framework and modeling approach classification for chemical production scheduling. *AIChE J.* **2012**, *58*, 1812–1828.
5. Velez, S.; Maravelias, C.T. Reformulations and branching methods for mixed-integer programming chemical production scheduling models. *Ind. Eng. Chem. Res.* **2013**, *52*, 3832–3841.
6. Wassick, J.M.; Ferrio, J. Extending the resource task network for industrial applications. *Comput. Chem. Eng.* **2011**, *35*, 2124–2140.
7. Nie, Y.; Biegler, L.T.; Villa, C.M.; Wassick, J.M. Discrete Time Formulation for the Integration of Scheduling and Dynamic Optimization. *Ind. Eng. Chem. Res.* **2015**, *54*, 4303–4315.
8. Gupta, D.; Maravelias, C.T.; Wassick, J.M. From rescheduling to online scheduling. *Chem. Eng. Res. Des.* **2016**, *116*, 83–97.
9. Cott, B.J.; Macchietto, S. Minimizing the effects of batch process variability using online schedule modification. *Comput. Chem. Eng.* **1989**, *13*, 105–113.
10. Kanakamedala, K.B.; Reklaitis, G.V.; Venkatasubramanian, V. Reactive schedule modification in multipurpose batch chemical plants. *Ind. Eng. Chem. Res.* **1994**, *33*, 77–90.
11. Huercio, A.; Espuña, A.; Puigjaner, L. Incorporating on-line scheduling strategies in integrated batch productioncontrol. *Comput. Chem. Eng.* **1995**, *19*, 609–614.
12. Kim, M.; Lee, I.B. Rule-based reactive rescheduling system for multi-purpose batch processes. *Comput. Chem. Eng.* **1997**, *21*, S1197–S1202.
13. Ko, D.; Na, S.; Moon, I.; Oh, M.; Dong-Gu Samsung, T.S. Development of a Rescheduling System for the Optimal Operation of Pipeless Plants. *Comput. Chem. Eng.* **1999**, *23*, S523–S526.
14. Huang, W.; Chung, P.W.H. A constraint approach for rescheduling batch processing plants including pipeless plants. *Comput. Aided Chem. Eng.* **2003**, *14*, 161–166.
15. Henning, G.P.; Cerdá, J. Knowledge-based predictive and reactive scheduling in industrial environments. *Comput. Chem. Eng.* **2000**, *24*, 2315–2338.
16. Palombarini, J.; Martínez, E. SmartGantt—An interactive system for generating and updating rescheduling knowledge using relational abstractions. *Comput. Chem. Eng.* **2012**, *47*, 202–216.
17. Elkamel, A.; Mohindra, A. A rolling horizon heuristic for reactive scheduling of batch process operations. *Eng. Optim.* **1999**, *31*, 763–792.
18. Vin, J.; Ierapetritou, M.G. A new approach for efficient rescheduling of multiproduct batch plants. *Ind. Eng. Chem. Res.* **2000**, *39*, 4228–4238.
19. Méndez, C.A.; Cerdá, J. Dynamic scheduling in multiproduct batch plants. *Comput. Chem. Eng.* **2003**, *27*, 1247–1259.
20. Ferrer-Nadal, S.; Méndez, C.A.; Graells, M.; Puigjaner, L. Optimal reactive scheduling of manufacturing plants with flexible batch recipes. *Ind. Eng. Chem. Res.* **2007**, *46*, 6273–6283.
21. Janak, S.L.; Floudas, C.A.; Kallrath, J.; Vormbrock, N. Production scheduling of a large-scale industrial batch plant. II. Reactive scheduling. *Ind. Eng. Chem. Res.* **2006**, *45*, 8253–8269.
22. Novas, J.M.; Henning, G.P. Reactive scheduling framework based on domain knowledge and constraint programming. *Comput. Chem. Eng.* **2010**, *34*, 2129–2148.
23. Honkomp, S.; Mockus, L.; Reklaitis, G.V. A framework for schedule evaluation with processing uncertainty. *Comput. Chem. Eng.* **1999**, *23*, 595–609.
24. Subramanian, K.; Maravelias, C.T.; Rawlings, J.B. A state-space model for chemical production scheduling. *Comput. Chem. Eng.* **2012**, *47*, 97–110.
25. Gupta, D.; Maravelias, C.T. On deterministic online scheduling: Major considerations, paradoxes and remedies. *Comput. Chem. Eng.* **2016**, *94*, 312–330.

26. Velez, S.; Maravelias, C.T. Advances in Mixed-Integer Programming Methods for Chemical Production Scheduling. *Annu. Rev. Chem. Biomol. Eng.* **2014**, *5*, 97–121.
27. Subramanian, K.; Rawlings, J.B.; Maravelias, C.T.; Flores-Cerrillo, J.; Megan, L. Integration of control theory and scheduling methods for supply chain management. *Comput. Chem. Eng.* **2013**, *51*, 4–20.
28. Subramanian, K.; Rawlings, J.B.; Maravelias, C.T. Economic model predictive control for inventory management in supply chains. *Comput. Chem. Eng.* **2014**, *64*, 71–80.
29. Kondili, E.; Pantelides, C.C.; Sargent, R.W.H. A general algorithm for short-term scheduling of batch operations-I. MILP formulation. *Comput. Chem. Eng.* **1993**, *17*, 211–227.
30. Pantelides, C.C. Unified frameworks for optimal process planning and scheduling. In *Proceedings of the Second Conference on Foundations of Computer Aided Operations*; Cache: New York, NY, USA, 1994; pp. 253–274.
31. Sundaramoorthy, A.; Maravelias, C.T. Computational Study of Network-Based Mixed-Integer Programming Approaches for Chemical Production Scheduling. *Ind. Eng. Chem. Res.* **2011**, *50*, 5023–5040.
32. Pinto, J.M.; Grossmann, I.E. A Continuous Time Mixed Integer Linear Programming Model for Short Term Scheduling of Multistage Batch Plants. *Ind. Eng. Chem. Res.* **1995**, *34*, 3037–3051.
33. Blomer, F.; Gunther, H.O. LP-based heuristics for scheduling chemical batch processes. *Ind. Eng. Chem. Res.* **2000**, *38*, 1029–1051.
34. Velez, S.; Maravelias, C.T. Mixed-integer programming model and tightening methods for scheduling in general chemical production environments. *Ind. Eng. Chem. Res.* **2013**, *52*, 3407–3423.
35. Merchan, A.F.; Maravelias, C.T. Reformulations of Mixed-Integer Programming Continuous-Time Models for Chemical Production Scheduling. *Ind. Eng. Chem. Res.* **2014**, *53*, 10155–10165.
36. Burkard, R.; Hatzl, J. Review, extensions and computational comparison of MILP formulations for scheduling of batch processes. *Comput. Chem. Eng.* **2005**, *29*, 1752–1769.
37. Janak, S.L.; Floudas, C.A. Improving unit-specific event based continuous-time approaches for batch processes: Integrality gap and task splitting. *Comput. Chem. Eng.* **2008**, *32*, 913–955.
38. Lee, H.; Maravelias, C.T. Discrete-time mixed-integer programming models for short-term scheduling in multipurpose environments. *Comput. Chem. Eng.* **2017**, *107*, 171–183.
39. Sahinidis, N.; Grossmann, I. Reformulation of multiperiod MILP models for planning and scheduling of chemical processes. *Comput. Chem. Eng.* **1991**, *15*, 255–272.
40. Yee, K.; Shah, N. Improving the efficiency of discrete time scheduling formulation. *Comput. Chem. Eng.* **1998**, *22*, S403–S410.
41. Lee, H.; Maravelias, C.T. Mixed-integer programming models for simultaneous batching and scheduling in multipurpose batch plants. *Comput. Chem. Eng.* **2017**, *106*, 621–644.
42. Papageorgiou, L.G.; Pantelides, C.C. Optimal campaign planning/scheduling of multipurpose batch/semicontinuous plants. 2. A mathematical decomposition approach. *Ind. Eng. Chem. Res.* **1996**, *35*, 510–529.
43. Bassett, M.H.; Pekny, J.F.; Reklaitis, G.V. Decomposition techniques for the solution of large-scale scheduling problems. *AIChE J.* **1996**, *42*, 3373–3387.
44. Kelly, J.D.; Zyngier, D. Hierarchical decomposition heuristic for scheduling: Coordinated reasoning for decentralized and distributed decision-making problems. *Comput. Chem. Eng.* **2008**, *32*, 2684–2705.
45. Wu, D.; Ierapetritou, M.G. Decomposition approaches for the efficient solution of short-term scheduling problems. *Comput. Chem. Eng.* **2003**, *27*, 1261–1276.
46. Calfa, B.A.; Agarwal, A.; Grossmann, I.E.; Wassick, J.M. Hybrid Bilevel-Lagrangean Decomposition Scheme for the Integration of Planning and Scheduling of a Network of Batch Plants. *Ind. Eng. Chem. Res.* **2013**, *52*, 2152–2167.
47. Castro, P.M.; Harjunkoski, I.; Grossmann, I.E. Greedy algorithm for scheduling batch plants with sequence-dependent changeovers. *AIChE J.* **2011**, *57*, 373–387.
48. Roslöf, J.; Harjunkoski, I.; Björkqvist, J.; Karlsson, S.; Westerlund, T. An MILP-based reordering algorithm for complex industrial scheduling and rescheduling. *Comput. Chem. Eng.* **2001**, *25*, 821–828.
49. Kopanos, G.M.; Méndez, C.A.; Puigjaner, L. MIP-based decomposition strategies for large-scale scheduling problems in multiproduct multistage batch plants: A benchmark scheduling problem of the pharmaceutical industry. *Eur. J. Oper. Res.* **2010**, *207*, 644–655.
50. Relvas, S.; Barbosa-Póvoa, A.P.F.; Matos, H.A. Heuristic batch sequencing on a multiproduct oil distribution system. *Comput. Chem. Eng.* **2009**, *33*, 712–730.

51. Jain, V.; Grossmann, I.E. Algorithms for Hybrid MILP/CP Models for a Class of Optimization Problems. *INFORMS J. Comput.* **2001**, *13*, 258–276.
52. Harjunkoski, I.; Grossmann, I.E. Decomposition techniques for multistage scheduling problems using mixed-integer and constraint programming methods. *Comput. Chem. Eng.* **2002**, *26*, 1533–1552.
53. Maravelias, C.T.; Grossmann, I.E. A hybrid MILP/CP decomposition approach for the continuous time scheduling of multipurpose batch plants. *Comput. Chem. Eng.* **2004**, *28*, 1921–1949.
54. Roe, B.; Papageorgiou, L.G.; Shah, N. A hybrid MILP/CLP algorithm for multipurpose batch process scheduling. *Comput. Chem. Eng.* **2005**, *29*, 1277–1291.
55. Maravelias, C.T. A decomposition framework for the scheduling of single- and multi-stage processes. *Comput. Chem. Eng.* **2006**, *30*, 407–420.
56. Subrahmanyam, S.; Kudva, G.K.; Bassett, M.H.; Pekny, J.F. Application of distributed computing to batch plant design and scheduling. *AIChE J.* **1996**, *42*, 1648–1661.
57. Ferris, M.C.; Maravelias, C.T.; Sundaramoorthy, A. Simultaneous Batching and Scheduling Using Dynamic Decomposition on a Grid. *INFORMS J. Comput.* **2009**, *21*, 398–410.
58. Velez, S.; Maravelias, C.T. A branch-and-bound algorithm for the solution of chemical production scheduling MIP models using parallel computing. *Comput. Chem. Eng.* **2013**, *55*, 28–39.
59. Shah, N.; Pantelides, C.C.; Sargent, R.W.H. A general algorithm for short-term scheduling of batch operations-II. Computational issues. *Comput. Chem. Eng.* **1993**, *17*, 229–244.
60. Stephanopoulos, G. *Chemical Process Control: An Introduction to Theory and Practice*; Prentice-Hall: Englewood Cliffs, NJ, USA, 1984; p. 696.
61. Ogunnaike, B.A.; Ray, W.H. *Process Dynamics, Modeling, and Control*; Oxford University Press: New York, NY, USA, 1994; p. 1260.
62. Bequette, B.W. *Process Control : Modeling, Design, and Simulation*; Prentice Hall PTR: Upper Saddle River, NJ, USA, 2003; p. 769.
63. Rawlings, J.B.; Mayne, D. *Model Predictive Control: Theory and Design*; Nob Hill Pub: Madison, WI, USA, 2009; p. 669.
64. Seborg, D.E.; Edgar, T.F.; Duncan, M.A.; Doyle, F.J., III. *Process Dynamics and Control*; Wiley: Hoboken, NJ, USA, 2016; p. 502.
65. Amrit, R.; Rawlings, J.B.; Biegler, L.T. Optimizing process economics online using model predictive control. *Comput. Chem. Eng.* **2013**, *58*, 334–343.
66. Ellis, M.; Durand, H.; Christofides, P.D. A tutorial review of economic model predictive control methods. *J. Process Control* **2014**, *24*, 1156–1178.
67. Rawlings, J.B.; Risbeck, M.J. Model predictive control with discrete actuators: Theory and application. *Automatica* **2017**, *78*, 258–265.
68. Baldea, M.; Harjunkoski, I. Integrated production scheduling and process control: A systematic review. *Comput. Chem. Eng.* **2014**, *71*, 377–390.
69. Li, Z.; Ierapetritou, M.G. Process scheduling under uncertainty: Review and challenges. *Comput. Chem. Eng.* **2008**, *32*, 715–727.
70. Janak, S.L.; Lin, X.; Floudas, C.A. A new robust optimization approach for scheduling under uncertainty. II. Uncertainty with known probability distribution. *Comput. Chem. Eng.* **2007**, *31*, 171–195.
71. Sand, G.; Engell, S. Modeling and solving real-time scheduling problems by stochastic integer programming. *Comput. Chem. Eng.* **2004**, *28*, 1087–1103.
72. Sabuncuoglu, I.; Karabuk, S. Rescheduling frequency in an fms with uncertain processing times and unreliable machines. *J. Manuf. Syst.* **1999**, *18*, 268–283.
73. Chaari, T.; Chaabane, S.; Aissani, N.; Trentesaux, D. Scheduling under uncertainty: Survey and research directions. In Proceedings of the 2014 International Conference on Advanced Logistics and Transport, Hammamet, Tunisia, 1–3 May 2014; pp. 229–234.
74. Martagan, T.; Krishnamurthy, A. Control and Optimization of Bioprocesses Using Markov Decision Process. In Proceedings of the 2012 Industrial and Systems Engineering Research Conference, Orlando, FL, USA, 19–23 May 2012; pp. 1–8.
75. Martagan, T.; Krishnamurthy, A.; Maravelias, C.T. Optimal condition-based harvesting policies for biomanufacturing operations with failure risks. *IIE Trans.* **2016**, *48*, 440–461.

76. Dedopoulos, I.T.; Shah, N. Optimal Short-Term Scheduling of Maintenance and Production for Multipurpose Plants. *Ind. Eng. Chem. Res.* **1995**, *34*, 192–201.
77. Sanmartí, E.; Espuña, A.; Puigjaner, L. Batch production and preventive maintenance scheduling under equipment failure uncertainty. *Comput. Chem. Eng.* **1997**, *21*, 1157–1168.
78. Vassiliadis, C.; Pistikopoulos, E. Maintenance scheduling and process optimization under uncertainty. *Comput. Chem. Eng.* **2001**, *25*, 217–236.
79. Kopanos, G.M.; Xenos, D.P.; Cicciotti, M.; Pistikopoulos, E.N.; Thornhill, N.F. Optimization of a network of compressors in parallel: Operational and maintenance planning – The air separation plant case. *Appl. Energy* **2015**, *146*, 453–470.
80. Xenos, D.P.; Kopanos, G.M.; Cicciotti, M.; Thornhill, N.F. Operational optimization of networks of compressors considering condition-based maintenance. *Comput. Chem. Eng.* **2016**, *84*, 117–131.
81. Biondi, M.; Sand, G.; Harjunkoski, I. Optimization of multipurpose process plant operations: A multi-time-scale maintenance and production scheduling approach. *Comput. Chem. Eng.* **2017**, *99*, 325–339.
82. Zhang, Y.H.P.; Sun, J.; Ma, Y. Biomanufacturing: History and perspective. *J. Ind. Microbiol. Biotechnol.* **2017**, *44*, 773–784.
83. Clomburg, J.M.; Crumbley, A.M.; Gonzalez, R. Industrial biomanufacturing: The future of chemical production. *Science* **2017**, *355*, doi:10.1126/science.aag0804.

processes

MDPI

Article

A Validated Model for Design and Evaluation of Control Architectures for a Continuous Tablet Compaction Process

Fernando Nunes de Barros, Aparajith Bhaskar and Ravendra Singh *

Engineering Research Center for Structured Organic Particulate Systems (ERC-SOPS),
Department of Chemical and Biochemical Engineering, Rutgers, The State University of New Jersey,
Piscataway, NJ 08854, USA; fernandondebarros@gmail.com (F.N.d.B.); aparajithbhaskar94@gmail.com (A.B.)
* Correspondence: ravendra.singh@rutgers.edu or ravendra_01@yahoo.com; Tel.: +1-848-445-4944

Received: 2 October 2017; Accepted: 28 November 2017; Published: 1 December 2017

Abstract: The systematic design of an advanced and efficient control strategy for controlling critical quality attributes of the tablet compaction operation is necessary to increase the robustness of a continuous pharmaceutical manufacturing process and for real time release. A process model plays a very important role to design, evaluate and tune the control system. However, much less attention has been made to develop a validated control relevant model for tablet compaction process that can be systematically applied for design, evaluation, tuning and thereby implementation of the control system. In this work, a dynamic tablet compaction model capable of predicting linear and nonlinear process responses has been successfully developed and validated. The nonlinear model is based on a series of transfer functions and static polynomial models. The model has been applied for control system design, tuning and evaluation and thereby facilitate the control system implementation into the pilot-plant with less time and resources. The best performing control algorithm was used in the implementation and evaluation of different strategies for control of tablet weight and breaking force. A characterization of the evaluated control strategies has been presented and can serve as a guideline for the selection of the adequate control strategy for a given tablet compaction setup. A strategy based on a multiple input multiple output (MIMO) model predictive controller (MPC), developed using the simulation environment, has been implemented in a tablet press unit, verifying the relevance of the simulation tool.

Keywords: tablet press; nonlinear model; model predictive control; continuous manufacturing; quality by control; critical quality attributes

---

1. Introduction

Due to the enablement of more reliable and faster drug production; the pharmaceutical industry in recent times is showing great interest in continuous manufacturing (CM) technology. There are many advantages of adapting such a technology over batch manufacturing. The increased speed of production can help manufacturers respond in a faster manner to changes in demand. Integrating quality by design (QbD), quality by control (QbC), real time release (RTR) and efficient process analytical technology (PAT) can serve to cut down manufacturing costs and improve drug quality and thereby patient safety. This integration of concepts to enable the transition from batch to continuous manufacturing requires intensive research in terms of understanding processes from a mechanistic perspective rather than a black box approach.

Quality by design, as introduced by the ICH Q8 guidance on pharmaceutical development, has required that companies demonstrate an understanding of how various elements in the process affect the product quality [1,2]. This would entail being able to correlate critical process parameters to critical quality attributes within each unit operation. The development of such an understanding can help formulate control strategies that can efficiently either constrain variables within their design space or track their set points. The development of control strategies experimentally can be taxing on material, time and is subject to equipment related and operating constraints. The dependency on these factors lays the grounds for a need to create a virtual environment on which experiments can be simulated. Not only would this speed up the process development but also improve the operational efficiency through a model derived better process understanding.

There have been multiple works in the past that model the integrated pharmaceutical manufacturing plant. The work by Mesbah et al. involves simulating chemical synthesis, purification, formulation and tableting from start to finish in a continuous mode [3]. Work done by Sen et al. focuses on the influence of upstream API properties on downstream quality metrics [4]. Boukouvala et al. focuses on identifying the challenges in flowsheet model development and simulation for solid-based pharmaceutical processes [5]. Though these works incorporate dynamics of the unit operation, important interactions between process variables is lacking. This work will focus on capturing these interactions within the tablet compaction unit operation, which can in turn be integrated into a larger flow sheet model. The systematic modelling framework described in this manuscript will help guide future modelling efforts for pharmaceutical process controls.

Another motivation for this work lies in the fact that, during manufacturing, tablet compaction is subject to disturbances in density and concentration. This in turn can lead to product that is of low quality. Developing a control system that can effectively control the critical quality attributes (CQAs) within the tablet process is an essential requirement to fully establishing a robust continuous manufacturing process. Previous studies contribute significantly through experimental and simulated work, with respect to understanding compaction in terms of material properties and formulations [6–8]. Significant work has been done to create an understanding of the interaction between process parameters [9]. Previous studies also involved the modelling aspects of the tablet press [7,10–13]. The work developed so far has either been purely empirical or mechanistic FEA (Finite Element Analysis) or DEM (Discrete Element Modelling) based simulation. Although contributing greatly in developing an understanding of compaction, none of these works were oriented towards developing a control system.

The optimization based control scheme, that is, model predictive control (MPC) is inherently capable of dealing with multi input multi output (MIMO) interactive and dead time dominant systems. It also allows the inclusion of process related constraints in its design. These advantages have been manifested in its wide spread usage [14–16]. In recent years, multiple variations of the same have been developed [17–20]. There have been developments in MPC schemes that include model uncertainty to improve its robustness [18]. Another approach to deal with model uncertainty is explored through an MPC design called a scenario based MPC [21]. There is also a significant amount of literature and development in nonlinear model predictive control [22]. Depending on the situation these strategies maybe used to improve the efficiency of the control system. MPC has been very successful in chemical industries. However, much less attention has been paid to develop and implement the MPC in pharmaceutical tablet manufacturing process.

In terms of the work based in the domain of process control, there have been some contributions to expansion of the knowledge base in the pharmaceutical industry. However, these works were separated into two independent segments i.e., simulation and experimental. The simulation work is based on theoretically derived equations [23–25]. Model predictive control has been implemented in a number of simulation studies [23,24,26]. Experimental work has been conducted for feeder and blender unit operations [27]. There has been some work done in the development of PAT (Process Analytical Technology) for the utilization in control strategies [28]. Based on this, work hybrid control strategies have been developed for continuous pharmaceutical manufacturing [28–31]. There has also been experimental work done using a commercial model predictive control platform for the continuous direct compaction process [32]. Despite the breadth of the control work in the pharmaceutical space, there is still a disconnect between the simulation and experimental environments. With the best of authors knowledge, no effort has been made to integrate the simulation studies on pharmaceutical process control involving solid dosages forms with experimental implementation works. Therefore, the commercialization of the concepts developed through simulation is still a challenging task.

In this work, a transfer function based model of the tablet compaction process has been developed to enable the design and evaluation of control systems. The non-linear process behavior and the dependency of process delays on process operating parameters have been considered. The model was validated against experimental data and it was demonstrated that the selected modelling strategy, which is based on compression ratios, is capable of representing the nonlinear behavior of the compaction forces. The applications of the model for control system design, tuning and evaluation has been demonstrated. The performance of three control algorithms was evaluated to determine the control algorithm to be used in the evaluation of different control strategies. Selected control strategies were chosen based on process understanding of the compaction process–critical process parameter interaction with critical quality attributes; and were implemented on this model. The set point tracking and disturbance rejection capabilities were tested, and the results are contextualized in terms of their usage. The developed model is postulated to be applicable to a wide range of situations, but this paper focuses on its application to design a control system before implementation into a real manufacturing set up. The screened control strategies are then implemented into the pilot-pant for experimental verification. A comparison of the experimental and simulated results is also summarized. Results show that the model was successfully derived and can be an asset in the development and evaluation of new control systems, by reducing the time and cost involved in this process.

In next section, the continuous pharmaceutical tablet manufacturing process has been described. Materials and methods used in this study has been summarized in Section 3. The tablet compaction process model has been developed and validated in Section 4. Applications of the model for control system design and evaluation has been described in Section 5 while the controllers have been developed and tuned in Section 6. The results have been presented in Section 7. Finally, the manuscript has been concluded in Section 8.

2. Continuous Pharmaceutical Tablet Manufacturing Process

2.1. Pilot-Plant

The continuous direct compaction tablet manufacturing pilot-plant that has been used for this study is shown in Figure 1. The pilot-plant is situated at ERC-SOPS, Rutgers University, USA [33]. The pilot-plant is built in three levels in such a way that gravity can be used as the driving force for material transport. The top most level is assigned to feeding and powder storage, the middle layer is designated to the task of blending and de-lumping and the bottom layer is used for compaction. Each of these levels spans an area of at least 10×10 square feet. The current equipment set includes three gravimetric feeders with the capability of expansion. Following the feeders, a co-mill is integrated for de-lumping the powders as mentioned before and creating contact between the components. The lubricant feeder is added after the co-mill to prevent over lubrication of the formulation in the co-mill. All these streams are then connected to a continuous blender to create a homogenous mixture of all the ingredients. The exit stream from the blender is fed to the tablet press via a rotary feed frame. The powder blend fills a die, which is subsequently compressed to create a tablet.

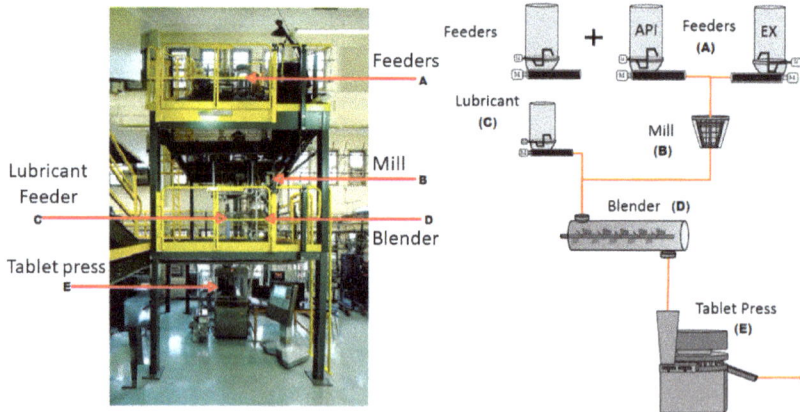

Figure 1. Direct compaction pilot plant.

2.2. Compaction Process

Tablet compaction takes place through a systematic series of steps. At the start of this process, the powder is fed into a rotary tablet press through a mechanical chute. The material then enters a feed frame where rotating blades fill powder into the dies one by one. An increase or decrease in the feed frame speed can change the instantaneous density through consolidation or can decrease the instantaneous density by fluidization depending on the speed and material properties. Although this is an adjustable parameter, its effect has not been explored in this paper. Feed frame speed was kept constant throughout the experiments since, from a control perspective, this parameter is not considered to be an actuator in any of the tablet compaction control strategies. The effect of this parameter on the compression forces is outside the scope of this work. Another important parameter is the fill depth. This defines the total volume of powder that will be filled at this stage through an adjustable height. An increase in this parameter is essentially an increase in the depth to which powder can be filled thus, increasing the volume of powder that is filled and subsequently the weight. The powder, once filled in the dies has the excess removed by the scraper. Thus, at this stage the weight of the tablet is defined.

After this, the powder goes through two compression stages—the pre-compression stage and the main compression stage. With respect to both stages, the mechanics of this is such that two rotating drums on the top and bottom can be adjusted in terms of the spacing between them. The drum at the bottom can be moved vertically to change the spacing, while the upper drum is kept stationary. At their respective stations, the spacing between the two drums determines the final height that the powder will be compacted to. The two parameters that can be adjusted are the pre-compression height and the main compression height. When the dies come in contact with the drums during rotation the top die presses down towards the bottom one thus, generating a compaction. The force the upper die experiences during this process is essentially the compression force data that can be extracted from the press and used for control. The pre-compression force station is necessary as it reduces phenomena such as capping, increases the dwell time and also causes de-aeration of the powder [34]. The main compression force station is where the actual compaction takes place. For a certain fill depth, a decrease in main and pre-compression height increases the force the upper drum experiences. Subsequently, this also increases the breaking force and density of the tablets.

In the following sections, the incorporation of these mechanisms into the simulation tool Simulink (Mathworks) will be explored. The application of various control strategies to create a set of control options for different production scenarios will be elaborated in subsequent sections.

3. Materials and Methods

All the simulations were conducted in the Matlab and Simulink environment. Parameter Estimation Toolbox, Control System Toolbox and Model Predictive Control Toolbox, which are built into Matlab, were used for model regression and control system design.

The experiments conducted in this work used a blend with a composition of 89% lactose monohydrate (excipient), 9% acetaminophen (API) and 1% magnesium stearate (lubricant). The blend was prepared using a Glatt batch blender run at 25 revolutions per minute (rpm) for 30 min with a layered loading order to ensure that thorough mixing was achieved. The capacity of each batch was of 7 kg. Multiple batches had to be prepared throughout the experiments. Experimental data was collected through DeltaV (Emerson), which was also used as the platform for the implementation of control strategies into the pilot-plant.

Actuators for the control strategies used in this work were determined through a literature review, process understanding of the tablet compaction operation and a sensitivity analysis. Experiments were conducted to develop a main compression force, pre-compression force and simultaneous main and pre-compression force controllers. In these experiments, the fill depth and main compression height variables were used as manipulating variables. Each closed loop experiment was preceded by open loop experiments. During these experiments, the production rate was also varied to understand changes in transport delay. This experimental data has been used to develop the model and control architecture of tablet press.

4. Model Development and Validation

4.1. Systematic Modelling Framework for Pharmaceutical Process Control System Design

A systematic framework, shown in Figure 2, has been developed to guide future pharmaceutical modelling efforts. The framework starts with a conceptual model that is developed based on previous process knowledge. In this step, critical process parameters, critical quality attributes and any other relevant intermediate variables are identified. A simplified sensitivity analysis can help determine which variables should be integrated in the model. Once the model has been conceptualized, its architecture is implemented in Simulink (Mathworks). The model structure is defined according to the interaction between different variables and whether dynamic data can be obtained for all the variables. This structure can be derived empirically or based on first principles, being based on transfer functions, state-space and non-dynamic models.

The adequate experiments have been performed in order to generate process data. The materials and methods used in the experiments is described in Section 3. The control platform's continuous historian automatically stores the raw data generated during the experiments. The raw data of an entire day is then imported to a spreadsheet, which can be read by Matlab. All the data preprocessing and organization, such as classifying the data by experiments and converting the timestamps to an adequate format is done in Matlab.

Each individual model is then regressed to closely fit the experimental data. The regression procedure varies according to the type of model implemented. This work followed two different procedures, one for the regression of linear dynamic models (transfer functions) and one for nonlinear steady state models (polynomial). The former entailed the estimation of transfer function parameters via Matlab's Parameter Estimation Toolbox, while the latter was based on the Matlab's polynomial fit functionality (polyfit). The models were then individually validated. If the fitting was not considered adequate the procedure was repeated. Once adequate individual fittings were achieved, an integrated model validation was conducted. This was done running a simulation where the inputs mimicked input data from a previous experiment. The simulated results were then compared to the experimental results and the fitness of the model was determined. If the model was rejected, the model structure was modified and the regression steps were repeated. A new DOE was executed if not enough data is available to support the updated model.

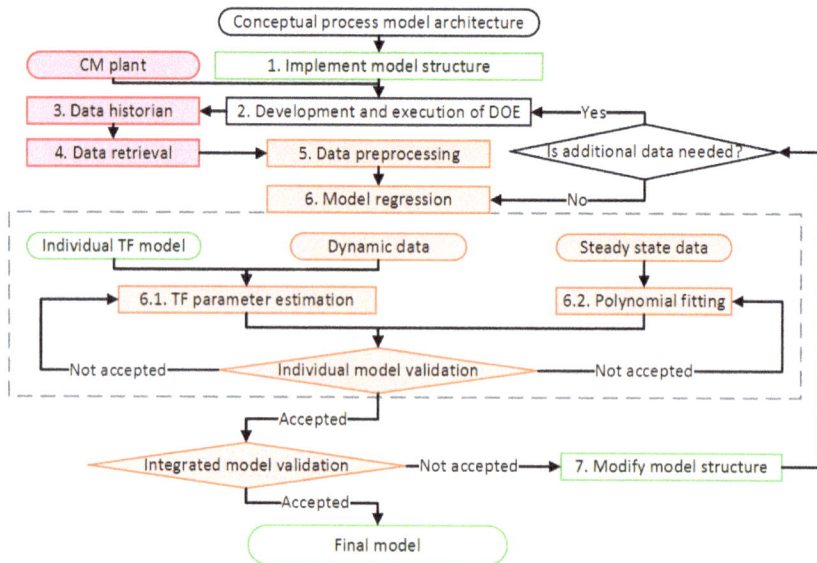

Figure 2. Systematic modelling framework. DOE: Design of experiments.

The accuracy of the models was quantified through the correlation coefficient (R^2). Once an accurate model was achieved ($R^2 > 0.9$), the model structure and its parameters were stored. The final model then has been used for the development and application of control systems.

4.2. Model Structure

4.2.1. Transfer Functions

Transfer functions are algebraic expressions that establish a dynamic relationship between an input and an output variable for a linear system [35]. The simulation model developed in this work consisted mostly of a series of empirical transfer functions of first and second order, each one representing a single input single output (SISO) system, arranged to capture the behavior of a multi input multi output system that is the tablet compaction equipment. The decision between first or second order transfer functions was based on the characteristic response of the process and on the regression coefficient (R^2) of the fitted model. A lower order transfer function was always preferred if its performance was satisfactory. Transfer functions are a compact manner of expressing a model and were selected as the main model structure due to their ease of implementation, interpretation and solution using Simulink. The form of a transfer function that was implemented in Simulink can be seen in Equation (1). A first order transfer function can be obtained from Equation (1) by setting the value of τ_1 to zero.

$$\frac{Y(s)}{U(s)} = \frac{K}{\tau_1 s^2 + \tau_2 s + 1} \tag{1}$$

where $U(s)$ and $Y(s)$ input and output signals, respectively, K is process gain and τ_i is the time constant. The values of the transfer function parameters are selected to minimize the difference between the simulated and experimental data. The fitting procedure is described in Section 4.4.1.

4.2.2. Variable Transport Delays

Multiple transport delays are present in the simulation model. From process understanding, it was learned that some of these transport delays varied as a function of the turret speed. The variable delays consist of two independent fractions as seen in Equation (2). The first fraction has a constant value and is inherent of the sensing method. The second fraction is dependent on the turret speed and the number of stations between the processes, varying according to Equation (3).

$$\theta = \theta_s + \theta_p\,(\omega) \tag{2}$$

$$\theta_p(\omega) = \frac{60\,\Delta n_p}{\omega\,n_p} \tag{3}$$

where θ_s is the sensor time delay, $\theta_p(\omega)$ is the process time delay, θ is the overall time delay, ω is turret speed, Δn_p is the number of punches between actuation and sensing and n_p is the total number of stations (punches).

4.3. Nonlinear Force Behavior

The main compression force and pre-compression force are the two main controlled variables as they directly affect the tablet breaking force and can be made to influence tablet weight. Previous literature provides information on the nonlinear correlation between tablet tensile strength and compaction force [36]. The Kawakita equation provides a correlation between the compaction pressure and volume [37]. In this work, a nonlinear correlation between compaction force and the ratio of the fill depth and the compaction height has been derived and presented. From the control perspective, this information is valuable as it gives direct information about the actuators and the controlled variables.

The experiments have been conducted to analyze the effect of compression ratio on main compression force. The compression ratio is defined as the ratio of fill depth and compression height. The steady state values of the force were plotted against their respective compression ratios. It is worth noting that the interactions between the fill depth and the compression forces are captured by relating the compression ratios to the compression forces. An increase in the fill depth and a decrease in the height would mean that more powder would have to be compressed to a smaller space, generating a higher force. A decrease in the fill depth and an increase in the height would result in the opposite scenario. Therefore, not only does this capture the dynamics of two actuators in one variable but it also assists in the modelling as it can be easily correlated to each of the forces.

From the plot (Figure 3), it was noticed that the steady state values of the force, as correlated with compression ratio, followed a nonlinear trend. Given a choice of an exponential equation and a second order polynomial equation, the latter provided a better fit to the experimental data. The form of the Equation (4) is shown below.

$$F = a_1 r^2 + a_2 r + a_3 \tag{4}$$

where r is ratio of fill depth and compression height, F is the compaction force and a_i are the polynomial coefficients of the model. The numerical values of these constants are provided in Section 4.6.2.

It is important to note that this model was obtained using purely the steady state data. Hence, this gives only non-dynamic information of the system. In order to capture the dynamics of the system, this polynomial can be preceded by a transfer function with a unit gain. The polynomial equation is a characteristic nonlinear equation.

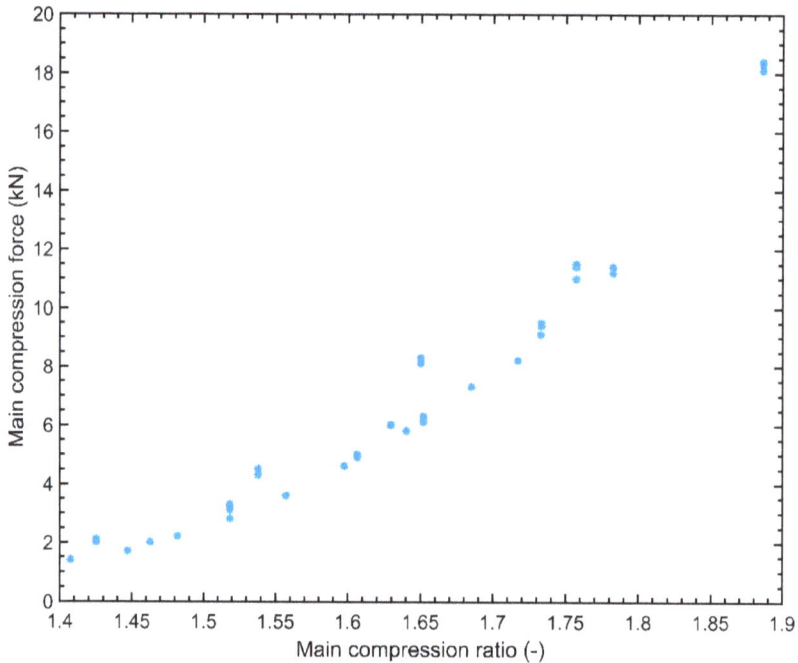

Figure 3. Main compression force (experimental data). Main compression ratio = FD/MCH.

4.4. Model Regression

4.4.1. Linear Models

Prior to the transfer function fitting it was necessary to determine, the critical process parameters and how they are affected by different process inputs. It has been achieved based on experiments and process knowledge. Once the inputs and outputs are paired for each individual transfer function, the fitting can take place.

The transfer function parameters were fitted using Matlab, Simulink and the Parameter Estimation Toolbox. Experimental input and output data for a determined transfer function is loaded to the Matlab workspace. A Simulink model consisting of a parameterized transfer function followed by a parameterized transport delay is then loaded and the Parameter Estimation Toolbox is used. In the Parameter Estimation Toolbox, the error between the experimental output and simulated output for a given input is minimized by varying the transfer function and transport delay parameters. The minimization is based on the optimization algorithm selected by the user. The cost function to be minimized can also be set by the user. All the transfer functions used in this paper were fitted using the nonlinear least squares methods with the Trust-Region-Reflexive algorithm. The sum of the squared errors was used as the cost function.

The linear models that were used to approximate the nonlinear dynamics of the compression forces were fitted in the following ranges: pre-compression force (1.4 kN–3.2 kN), main compression force (5 kN–12 kN). This linearization can be justified by the fact that these models are operated within a narrow range where this assumption is valid.

4.4.2. Nonlinear Models

The nonlinear models consisted a transfer function with unitary gain followed by a second order polynomial as described in Section 4.3. The polynomial was fitted based on steady state data using the polynomial fitting function, polyfit, built into Matlab. Initially, the possibility of capturing the behavior of both pre- and main compression forces using a single polynomial was evaluated. This single polynomial did not successfully fit the data. An alternative approach, where pre- and main compression forces are modelled by two different polynomials successfully fitted the data. A correlation between breaking force and main compression ratio was also modelled by a second order polynomial equation.

4.5. Model Implementation

A flexible process model was developed (using Simulink) to allow virtual experiments to be done in a quick and simple manner. The key input and output parameters of the tableting process are available for manipulation and monitoring in the model. The structure of the model was developed in such a way that the user can access each individual step of the tableting process and consists of four main modules. The first four modules capture the dynamic behavior of the mechanical actuators, pre-compression force, main compression force and tablet CQAs. The last module represents the tablet rejection system. The system is organized using Subsystem masks to facilitate the understanding of the implementation. An overview of the model implementation is presented in Figure 4. Three main modules and a snapshot of their implementation can be seen in this image.

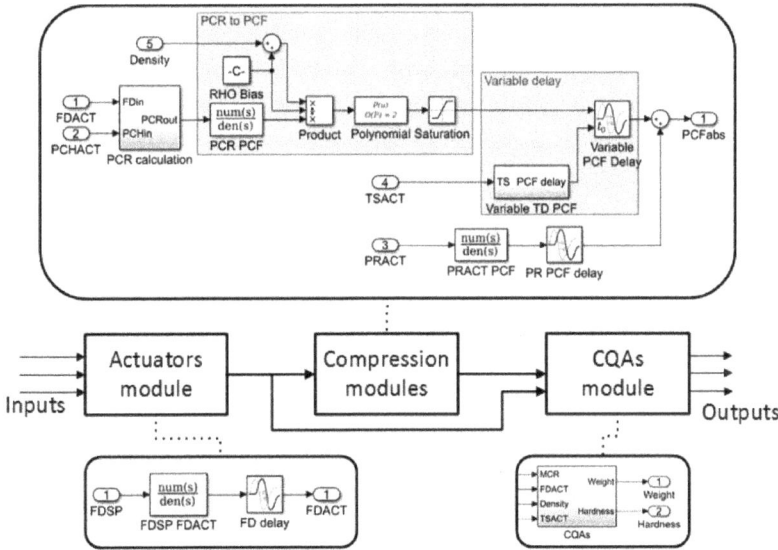

Figure 4. Model implementation overview. FDACT: Fill depth; PCR: Pre-compression ratio; PCF: Pre-compression force; RHO: Density; PRACT and PR: Production rate; TD: Transport delay; FDSP: Fill depth set point; CQA: Critical quality attribute.

4.5.1. Actuators Module

The actuator module consists of a series of transfer functions relating the set point and actual values for the actuators. A first order dynamic behavior was considered to represent the fill depth, pre-compression height and main compression height actuators. A second order transfer function was used to represent the production rate actuator. In this module turret speed is also calculated based on the actual production rate (PR) value using Equation (5).

$$\omega = \frac{1000PR}{60N_p} \tag{5}$$

where turret speed (ω) is given in rotations per minute, production rate is given in thousands of tablets per hour and N_p is the total number of stations (punches) in the press.

4.5.2. Compression Force Modules

In order to effectively represent the behavior of the compression forces, its interaction with fill depth, compression heights and production rate were modeled. From process understanding, it was recognized that the compression force is affected by a combination of the amount of powder filled into the die and the height to which this powder is compressed. This interaction was captured by relating the ratio of fill depth and compression height (compression ratio) to the compression force. This model consisted of the nonlinear model that is described in Sections 4.3 and 4.4.2. The interaction between the production rate and the compression force was modelled using a second order transfer function and a constant transport delay. The response from the production rate model was added to the response from the nonlinear compression model. This signal is considered to be the final force value. It is important to note that the polynomial that was used in the nonlinear compression was fitted against the absolute values of the compression ratio. Since the variables in the module exist as deviation variables, it is appropriate to convert these variables into absolute values before using them as an input in the polynomial. Another important feature that was added to this module was a methodology to capture variations in the density. In the actual tablet press, a change in the powder density would result in an increased or decreased amount of powder being filled into the die. This behavior can be represented as linear changes in fill depth. The modified fill depth calculation is given in Equation (6).

$$FD^* = \frac{\rho}{\rho_{ref}} FD \tag{6}$$

where FD^* is the modified fill depth, FD is the actual fill depth, ρ is the powder bulk density in the feed frame and ρ_{ref} is the reference value of bulk density at which the polynomial coefficients of Equation (5) were fitted.

4.5.3. Critical Quality Attributes Module

The critical quality attributes module uses fill depth, density, compression forces and turret speed to calculate the tablet weight and breaking force. A first order transfer function with unitary gain followed by a weight calculation (Equation (7)) and a variable transport delay is applied to model the tablet weight behavior. Tablet breaking force is calculated through a polynomial relationship between tablet breaking force and main compression ratio followed by a transport delay. Changes in density have a direct effect on both tablet weight and breaking force. The effect of density changes in tablet breaking force was modeled using the modified fill depth approach shown in the previous section. It is important to note that the tablet weight transfer function models the characteristic behavior of the weight sensing technique used in the experiments, which is coupled with the fill depth dynamics to represent the overall process dynamics of tablet weight. A transfer function is not used to model tablet breaking force since no dynamic data was available for this variable because of sensing limitations. The polynomial equation used to model tablet breaking force is based solely on steady state measurements, which, when coupled with the fill depth dynamic model can represent the breaking force dynamic behavior. A fixed transport delay is used to represent a hypothetical sensor behavior.

$$W = A_p \rho FD \tag{7}$$

where W is tablet weight, A_p is the area of a tablet punch, ρ is the powder density and FD is the fill depth.

4.5.4. Tablet Rejection Module

A module representing the tablet rejection system present in the tablet press was also developed. This module quantifies the total amount of tablet produces during the simulation as well as quantifying the number of tablet inside and outside specification. The total production is calculated by an integrator function that takes the production rate as an input. The good production is determined by multiplying the production rate signal by a series of logic signals coming from relay blocks and then integrating the resulting signal over time. The production is classified as good production if all the relays send a true signal. Each relay block represents a specific quality attribute and if the tablets are not within the specifications for that quality attribute the relay blocks sends a false signal (zero). If the tablet is within specifications, the relay sends a true signal (one). The bad production is calculated by simply subtracting the good production from the total production.

4.6. Model Validation

Through mathematical regression as described in Section 4.4, the dynamics of the interaction between the various parameters were captured in mathematical models. A key feature of this modelling methodology is the importance to detail in terms of capturing all the process dynamics. The interaction between the various critical process parameters was captured. The dynamic between the set point and actual values was also identified. This section is organized in terms of the variables where a dynamic existed between set point and actual, critical process parameter and a respective actuator and critical quality attribute and critical process parameter.

4.6.1. Set Point and Actual Response Dynamics

The models have been developed to capture the dynamics between an operator providing a set point and this value being tracked by the final control element.

In Figure 5a, the simulated, experimental and set point values are plotted for fill depth. This model is driven by a transfer function and the regressed parameters for the same are provided in Table 1. This model is characterized by a response that is of first order dynamics. The figure shows step changes going from 6.15 mm to 5.7 mm in a step down and from 5.7 to 6.15 mm in a step up. These step changes are made in arbitrary time intervals to make sure of the model robustness. The close correlation between the simulated and experimental fill depth along with an R^2 value of 0.9966 indicates that the model is very accurate.

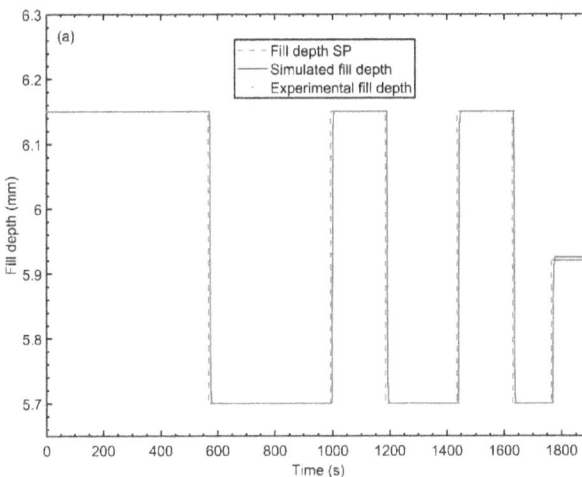

Figure 5. *Cont.*

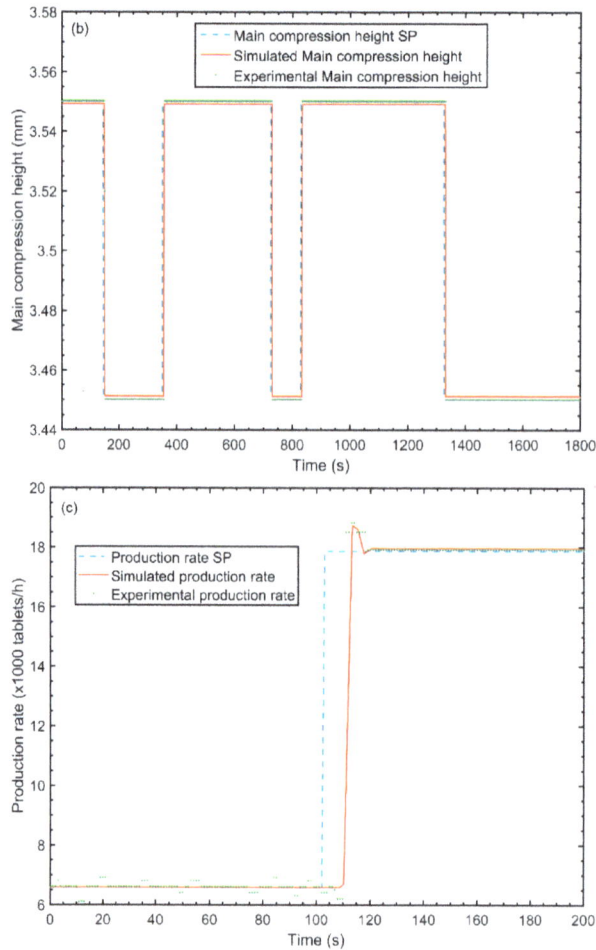

Figure 5. (**a**) Fill depth model validation (**b**) Main compression height model validation (**c**) Production rate model validation (SP: set point).

Table 1. Model parameters—Actuators.

Model Inputs	Model Outputs	Model Details					
		Order	τ_1 (s)	τ_2 (s)	θ (s)	Gain	R^2
Fill depth SP	Fill depth	1	1.0694	-	5.4986	1	0.9966
Main compression height SP	Main compression height	1	0.1658	-	5.3616	1	0.9973
Pre-compression height SP	Pre-compression height	1	0.1658	-	5.3616	1	0.9973
Production rate SP	Production rate	2	0.9	0.9968	8	1	0.9824

SP: Set point.

Figure 5b shows the model validation plot for the main compression height. It is similar to the fill depth dynamics in that it also follows a response characterized by a first order dynamic. Table 1 shows the regressed parameters that defined this transfer function. The figure describes the validation methodology. Step changes in the set point were provided and the experimental and simulated values were observed. The range of the step change is from 3.55 mm to 3.45 mm. As in the case of the fill depth, the step changes are made in arbitrary time intervals. This model is concluded to be accurate based on the R^2 value (0.9973) and the graph. The same model has been used to calculate the pre-compression height by assuming that it followed the same dynamics as the main compression height. The reason that this assumption holds good is that both the pre- and the main compression height essentially perform the same action of moving a drum to adjust compaction height. The difference lies in the fact that in all cases the final compaction height is lesser than the pre-compaction height. It was presumed that this difference in magnitude should not affect the dynamics of the model significantly.

The production rate model validation is shown in Figure 5c. This model is characterized by a response that follows a second order dynamic. The regressed parameters for the same are provided in Table 1. This figure shows one step change and the simulated and experimental responses. As can be seen, a slight overshoot is observed which is followed by an immediate convergence to the set point. This is characteristic of second order dynamics. The model accurately captures this as seen in the graph. The R^2 value (0.9824) is observed to be lower than the other models described in this section. This statistic is indicative of the inability of the model to capture slight variations that are observed in the graph.

4.6.2. Critical Process Parameter Interaction

As mentioned in Section 4.5.2, compression force modules were developed to capture the effects of the fill depth, pre and main compression heights, production rate and density on the pre- and main compression forces. Table 2 summarizes the details and parameters of the transfer function model for critical process parameters. Table 3 presents the polynomial coefficients from Equation (5) fitted to model the nonlinear behavior of the compression forces.

Table 2. Model parameters—Critical process parameters.

Model Inputs	Model Outputs	Model Details					
		Order	$\tau_1(s)$	$\tau_2(s)$	$\Theta max(s)$	Gain	R^2
Pre-compression ratio	Pre-compression force	1	2.5058	-	12.5	1	0.9680
Main compression ratio	Main compression force	1	3.4244	-	15	1	0.9659

Table 3. Polynomial coefficients.

Polynomial Constants	Pre-Compression Force (−)	Main Compression Force (−)
a_1	80.92	55.97
a_2	−219.40	−150.34
a_3	149.83	101.98

Figure 6a is a plot showing the model validation for the pre-compression force. The plot is divided into two levels. The top portion of the image shows changes in the pre-compression ratio, cause by variations in fill depth, while the bottom portion shows how these changes affect the pre-compression force in both the simulations and the experiments. Fill depth variations ranged from 6.6 mm to 5.8 mm. As observed in the image, there is a close matching between the developed model and the experimental values.

In Figure 6b, main compression force data, through both simulation and experiment, was plotted with respect to changes in main compression ratio caused by variations in both the fill depth and the main compression height. This model was developed to accommodate interaction in both its immediate actuators, that is, fill depth and main compression height. The image shows an increase in force for a decrease in the compaction height and a decrease in the force for a decrease in the fill depth. This is as expected and follows the same trend as the experimental data. The mismatches that are observed in the simulated signal's main compression force are captured in the lower R^2. This value is still considered to be high enough to use for control system development.

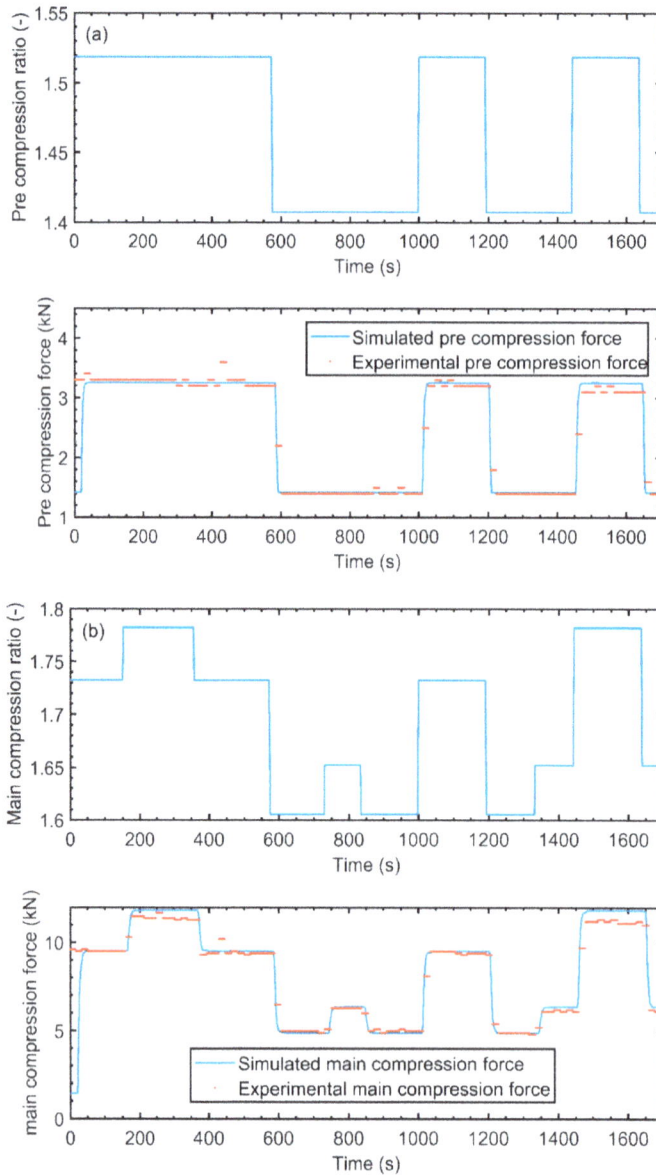

Figure 6. (a) Pre-compression force model validation (b) Main compression force model validation.

4.6.3. Critical Quality Attributes

As mentioned in Section 4.6.3, an attempt was made to model the influence of the critical process parameter's influence on the critical quality attributes. In this attempt, the tablet weight and breaking force models were developed. These have been validated in this section.

With respect to the tablet breaking force model, Figure 7a shows a graph plotting the breaking force versus the main compression ratio. It was previously mentioned that this model was developed based on steady state data. It is for this reason that the model was validated against more steady state data. As seen in the image the model used a second order polynomial to follow the nonlinear trend that is observed with the experimental data. The parameters for the polynomial that correlated main compression ratio and the tablet breaking force are $a_1 = 258.8846$, $a_2 = -695.3997$ and $a_3 = 468.2229$. This model was considered adequate for control system design. This section of the modelling is expected to be the focus of future research.

In Figure 7b, the model that was developed to predict continuous tablet weight was validated. Here, it is observed that when the fill depth is stepped down the experimental data follows a downward trend and stabilizes. The tablet weight reduces with a decrease in the fill depth. This experimental data follows the expected trend. Through regression, it was found that the downward dynamic trend could be captured using a first order transfer function. The parameters that were used to predict the dynamics of the tablet weight in the transfer function were a unit gain and a first order time constant of 6.5 s and a transport delay of 12 s.

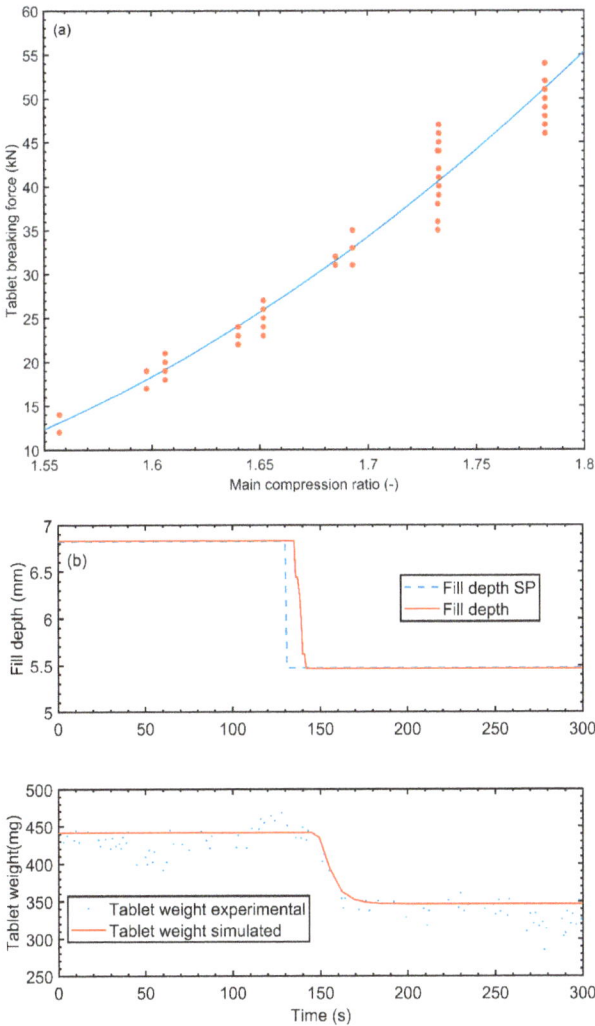

Figure 7. (a) Tablet breaking force model validation (b) Tablet weight model validation.

5. Application of the Developed Model for Design and Evaluation of Control Systems

Designing a new control strategy can be time and cost intensive. The high material cost and the limited material availability, especially during process development phase, can be a barrier for the implementation of control systems in the pharmaceutical industry. The developed model enables extensive experiments to be conducted on a virtual tablet press unit with little to no cost, generating valuable insights about the process. The ease of experimentation provided by the model allows for in-depth dynamic sensitivity analyses. Different manipulated-controlled variable pairings can be explored, since interaction between process variables are captured by the model. The performance of different control algorithms and strategies on this specific system can be easily studied. More complex evaluations of control systems, such as stability analysis through root locus diagrams and frequency response analysis, can also be derived from this model.

By integrating the developed model with the control platform through OPC connection it is possible to ensure that all the developed control modules have no flaws. A preliminary tuning of the implemented controllers can also be achieved through simulation, reducing the number of experiments necessary to deploy a new control strategy.

5.1. Control Algorithm Comparison

The set point tracking ability of three different control algorithms, PI, MPC and MPC with unmeasured disturbance model (integrated white noise model), was compared to determine which algorithm was the most adequate to be used for the development of the control strategies studied in this paper. The MPC algorithm with an unmeasured disturbance model consists of a variation of the conventional MPC, which removes steady state errors caused by unmeasured disturbances and mismatches between the controller model and the actual plant. A control strategy which controls main compression force through fill depth was used for the comparison due to its wide applicability in commercial tablet presses. Most of the commercially available tablet press have an inbuilt control system for main compression force but none of them has employed MPC algorithm.

The control algorithm comparison was conducted in the DCS (DeltaV Control Studio) and the simulation platform (Matlab/Simulink). The goal of this multiplatform implantation was to evaluate which control algorithm presents superior performance and to check if this superiority is observed in both platforms. A few differences can be pointed out between the controller implementations in each platform. The first difference is that features like anti-windup and manipulated variable tracking (during open loop) are set-up by default in the DCS, whereas in Simulink these features must be manually set up by the user, which may cause confusion if the user is transitioning from the DCS to Simulink. A difference in the PI controller implementation lies in the fact that the DCS and Simulink have slightly different forms for the PI equation, requiring a conversion of parameters to transfer tuning parameters between the two platforms. The MPC implementation also differs slightly between the two platforms. In the DCS the user does not have the flexibility to change the optimization algorithm used in the MPC. The models that can be used by the MPC in the DCS have to be either transfer function models or finite impulse response models (FIR), whereas in Simulink, the models can be either transfer function models or state-space models.

5.2. Control Strategy Development and Closed Loop Simulation

Three different novel control strategies were developed and evaluated as part of this study. The strategies were evaluated under two different scenarios. The first scenario consisted of a set point tracking experiment. The second scenario consisted of a disturbance rejection experiment, where variations in powder density were used as a disturbance. The variation in density used in the second scenario had the form of white noise disturbance, followed by ramp disturbance, succeeded by a step change in density (Figure 8).

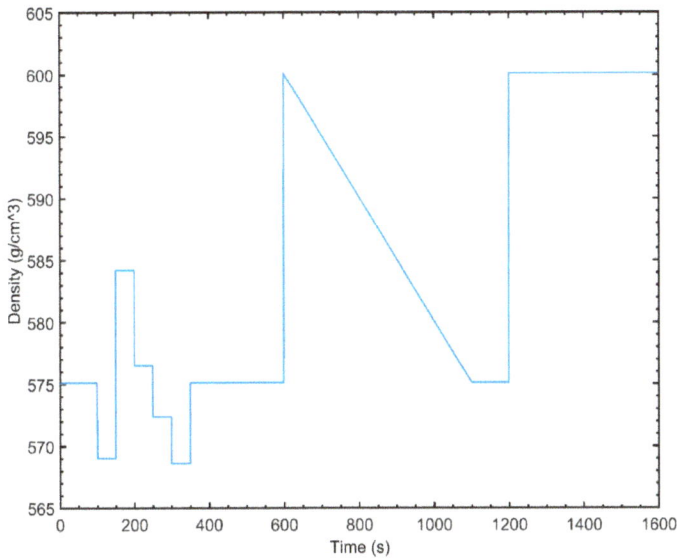

Figure 8. Density variations for disturbance rejection.

5.2.1. Strategy 1—Simultaneous Control of Pre- and Main Compression Forces

A control strategy for simultaneous control of pre- and main compression forces through fill depth and main compression height, respectively, using a MIMO MPC has been evaluated. Through a review of the compaction process, it is postulated that controlling pre-compression force by manipulating fill depth gives indirect control over the tablet weight and control of the main compression force through manipulations in main compression height gives indirect control of the tablet breaking force.

This was the only strategy that was also implemented in the pilot plant. This implementation was used to validate the applicability of the developed model. Currently, it is not feasible to measure tablet CQAs in real time with the available state of the art technology, making it impossible to implement strategies 2 and 3 in the pilot plant.

5.2.2. Strategy 2—Cascade Control of Weight and Breaking Force

Under circumstances where the sensors for both weight and breaking force are available, it is possible to use a cascade control strategy where the slave controller manipulates the fill depth and main compression height to control the pre- and main compression force respectively. The master controller manipulates the set points of the pre- and main compression forces to track the set points provided for weight and breaking force. This strategy's performance is dependent on the sampling rate of the sensors. A cascade control strategy works best when the slave controller is faster than the master controller. Given that available sensors for weight and tablet breaking force are slow due to their low sampling rates this control strategy is justified in its implementation.

5.2.3. Strategy 3—Direct Control of Tablet Weight and Cascade Tablet Breaking Force

A strategy where tablet weight is controlled directly through fill depth and tablet breaking force is controller through a cascade arrangement might be desired in when a fast sensor for tablet weight is available. Such sensor has been developed and implemented as a proof of concept in a previous work [32]. The strategy consists of a SISO main compression force controller, which manipulates the ratio between fill depth and main compression height. This ratio is used to calculate the value of main compression height while minimizing the interactions between fill depth and main compression force.

A secondary MPC is used to control tablet weight and breaking force through fill depth and main compression force set point.

6. Controller Tuning and Implementation

6.1. PI Controller

6.1.1. Tuning Procedure

The PI controller used in this work was tuned according to the modified SIMC method introduced by Skogestad (2011). This method consists of a rule based approach and considers both controller performance and robustness. An advantage of the SIMC method is that only a single tuning parameter (τ_c) needs to be adjusted. Lower τ_c values lead to a tighter controller (better performance), whereas higher values of τ_c lead to a smoother (more robust) controller. The value of $\tau_c = 20.03 \cong \theta$ was used to tune the PI controller, leading to values of $K_c = 0.0204$ and $\tau_I = 0.0019$. This value of τ_c is recommended by Skogestad as the tightest value that maintains a smooth control.

6.1.2. Implementation of PI Controller

The steps that were necessary to incorporate the PID block within the Simulink environment are elaborated in this section. A switch which closed the control loop was implemented before the PID block. This was done to replicate the production environment where the manufacturing line would be run in open loop during startup to allow for parameters to reach steady state before the controller is switched on. The signal from the switch is sent to the simulated plant and to the TR (signal tracking) slot on the PID block. During open loop operation, the signal received by the TR port serves to cancel out any action taken by the controller during this period. The output of the PID is constrained according to the required control strategy. The anti-windup method is set to back-calculation. This is done to avoid the saturation of the integral action when the output of the controller is constrained. It is important to note the PID block in Simulink uses P, I and D parameters, which need to be converted from the traditional K_c, τ_I and τ_D parameters calculated through the tuning method described in Section 6.1.1.

6.2. MPC

6.2.1. Tuning Procedure

The MPC makes use of an available process model to make predictions of future states of the system. An important facet of this process is the generation of a plant model that represents the process behavior accurately within the operational range. The process of developing a controller model for the implementation on the developed tablet press model is unique and can be summarized as follows.

The two main critical process parameters, pre-compression force and main compression force, used as controlled variables are nonlinear in nature and this nonlinearity is characterized by the gain of the system as described in Section 4.3. To be able to control this using a linear MPC, it is required to develop a linear model for the same. It is observed that on differentiating Equation (4), with respect to the ratio at a specific value, one arrives at the gain of the system at that operational point (Equation (8)).

$$k(r) = \left[\frac{d}{dr} (a_1 r^2 + a_2 r + a_3) \right] at \ r = R \tag{8}$$

where R is the desired value of ratio and k is the gain of the system.

Subsequently, this gain can be used with the previously regressed time constants to arrive at a linearized model of the system at a specific value of ratio. The nonlinear model is then replaced by this linear model and the order of the Pade approximation for all the time delay blocks are set to 30. The inbuilt MPC controller design feature is then used to generate a model. In this case, we proceed by saving the model in the Matlab workspace. The nonlinear model that originally existed is put back into place and the generated MPC model is then loaded into the controller. Once the nominal values of the inputs and outputs are defined, the controller is tuned. During tuning the values of control horizon, prediction horizon, penalty on move and penalty on error are defined. Through controller testing it was found that this method worked well but an offset in the final value persisted. This problem was resolved by incorporating an integrated white noise output disturbance model to the controller. Table 4 presents the tuning parameters for the MPC controllers used in this work.

Table 4. MPC tuning parameters.

| Controller | Ts (s) | P (Samples) | M (Samples) | Output Weight | | Input Rate Weight | |
				CV 1 (-)	CV 2 (-)	MV 1 (-)	MV 2 (-)
MCF	1	40	2	1	-	0.1	-
Strategy 1	1	25	2	0.135	0.135	0.739	0.739
Strategy 2—Master	4	20	2	1	1	0.1	0.1
Strategy 2—Slave	1	25	2	0.135	0.135	0.739	0.739
Strategy 3—Master	4	20	2	1	1	0.1	0.1
Strategy 3—Slave	4	20	2	1	-	0.1	-

Ts: Sampling time, P: Prediction horizon, M: Control horizon CV: Controlled variable, MV: Manipulated variable.

6.2.2. Implementation of MPC

Some key configurations were necessary to incorporate into the MPC block within the Simulink environment. A switch was implemented on the MPC block for the same as reason as in the case of the PI controller. In this case, the switch signal is sent to the MPC block through the mv target slot. The Timé delay blocks were approximated internally by the choosing a high value of 30 of the Pade order. This is so done because the model generation feature is not capable of dealing with nonlinearities. The Pade approximation linearizes the exponential time delay function.

7. Results

This section elaborates on the application of the developed model to the design of control systems. This simulation environment provides flexibility in terms of the different control strategies that can be implemented, the speed to experimentation and the lack of any material needs. First, control algorithms are compared in order to determine the best alternative. Then the selected control algorithm is used in the implementation of various strategies. Individual characteristics of the strategies are then discussed in order to guide the selection of the best control strategy of a given system.

7.1. Control Algorithm Performance Analysis

Two main compression force set point changes were applied to the system and its dynamic response was observed. The first step change was within the range at which the controllers were tuned and increased the MCF set point from 8 kN to 12 kN. The latter step change varied the MCF set point from 12 kN to 4 kN and was outside the range at which the controllers were tuned. The magnitude and range of the step changes were chosen in order to observe how linear controllers behave in different regions of a non-linear process.

Figure 9 shows the dynamic behavior of the different control algorithms in response to changes in the set point of the controlled variable.

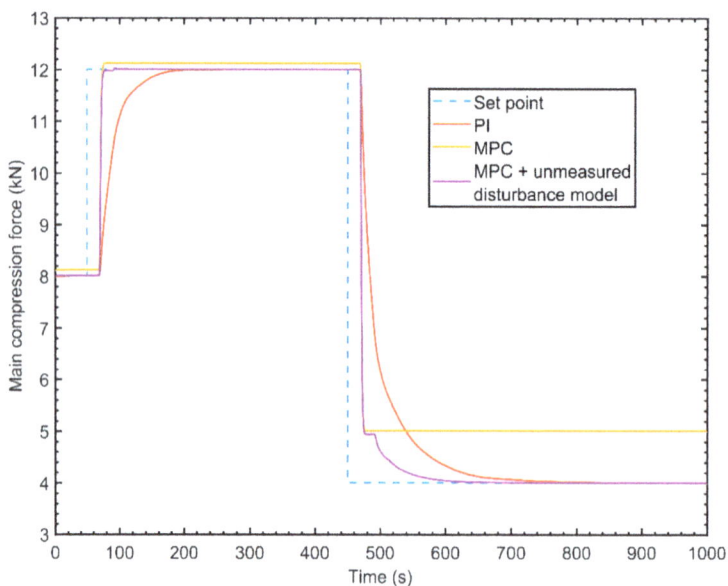

Figure 9. Control algorithm comparison. PI: Proportional Integral controller; MPC: Model predictive control.

From Figure 9, the percent overshoot and time to steady state are different for different step changes. This is expected since the linear controller are being implemented on a non-linear system. All three controllers presented a better performance when operated within the range at which they were tuned. A significant performance improvement can be noticed when an integrated white noise unmeasured disturbance model is added to the MPC controller. The unmeasured disturbance model can approximate mismatches between the linear MPC model and the non-linear behavior of the plant, eliminating the steady state error that is seen on the standard MPC controller. These mismatches become more significant as the set point moves away from the MPC ideal operational range. When a comparison between the PI controller and the MPC with unmeasured disturbance algorithms is made, the latter performs better than the former, giving a faster response with less overshoot.

The simulated experiments conducted using the model were replicated in pilot plant using the PI and MPC control algorithms available in the control platform (DeltaV). For this comparison, only the first step change was considered. Figure 10 shows both experimental and simulated results.

The superiority of MPC over the PI algorithm can also be seen in the experiments. A slight difference between the simulated and experimental results for similar control algorithms can be noticed in Figure 10. This difference is expected, since the controllers used in the experiments had a different tuning than the simulated controllers. The simulation environment allows for a better performing tuning to be achieved since there is no cost involved in iterating over different tuning parameters in order to find the optimal tuning.

Table 5 presents performance metrics for both simulated and experimental control algorithms. The MPC control algorithm will be used in all the following simulations conducted in this paper, since it is proven superior when compared to a PI algorithm.

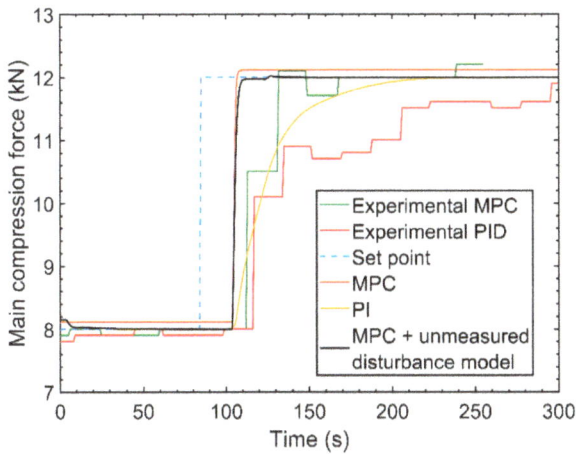

Figure 10. Simulated and actual closed loop results compared using a step change.

Table 5. Closed loop performance metrics.

Strategy	IAE (kN s)	ITAE (kN s)	ISE (kN s)	Rise Time (s)	Settling Time (s)	Overshoot (%)	Steady State Error (%)
Experimental PI	289.95	18506	693.45	121	211	0	0
Simulated PI	204.00	6596.2	626.48	64	120	6.11	0
Experimental MPC	180.50	5688.6	564.33	46	153	0	−7.50
Simulated MPC	111.65	5216.2	317.78	23	24	0	2.788
Simulated MPC with disturbance model	86.830	954.490	333.68	23	25	0.448	0

IAE: Integral absolute error, ITAE: Integral time-weighted absolute error, ISE: Integral squared error.

7.2. Control Strategy Evaluation

7.2.1. Strategy 1—Simultaneous Control of Pre- and Main Compression Forces

Figure 11a shows the dynamic response of the simulated system when set point changes are applied to both compression forces. The actuation signals are shown in Figure 11b. The implemented controller enables both variables to successfully track their respective set points. As expected, changes in main compression force set point have no effect on pre-compression force, since the main compression height (MCF actuator) does not interact with the pre-compression force. In an open loop scenario, manipulations in fill depth (PCF actuator) would lead to large variations in both forces. However, when the controller is on, changes in pre-compression force set point, which cause an actuation on fill depth, generate only a small deviation in main compression force. This occurs because the controller takes preventive actions on main compression height to mitigate the effect that fill depth changes have on main compression force. This indicates that the controller internal model is able to capture the interactions between manipulated and controller variables.

The effect that changes in the compression forces set points have on tablet weight and breaking force is shown in Figure 11c. Understanding the interaction between compression forces and tablet CQAs is necessary in order to achieve indirect control over them. No cross interaction between tablet weight and main compression force as well as tablet breaking force and pre-compression force is seen on the image, which indicates that tablet CQAs can be indirectly and independently controlled through the compression forces' set points. These set points can be determined manually by the operator or remotely by a supervisory controller, depending on the real-time data availability for tablet CQAs.

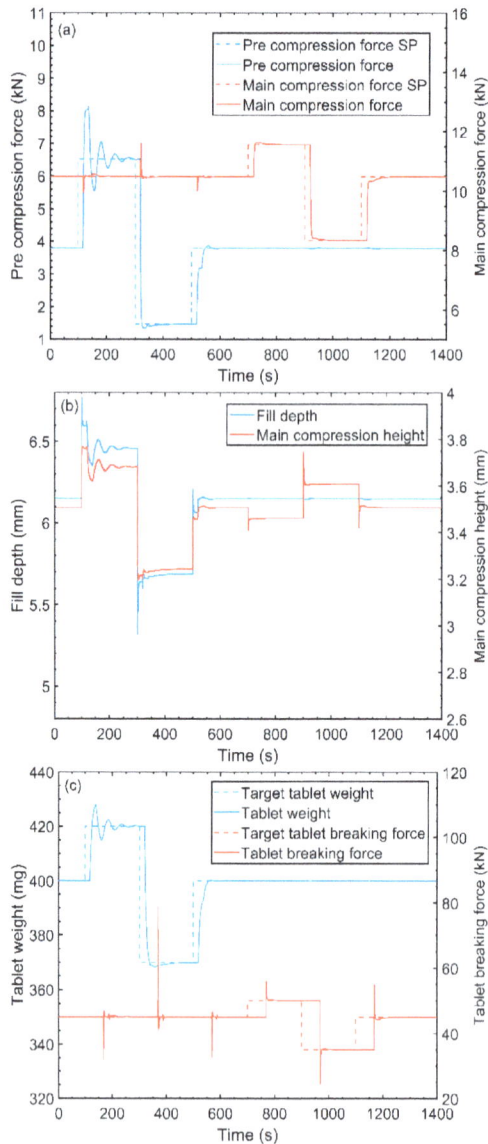

Figure 11. Control strategy 1—Set point tracking scenario. (**a**) Critical process parameters (**b**) Actuators (**c**) Critical quality attributes.

The system response to the disturbance rejection scenario is shown in Figure 12. When exposed to a white noise disturbance in density the controller takes action in order to bring the controlled variables back to their set points as seen in Figure 12a,b. The effect of the density variations is not fully mitigated by the controller due to the relatively high frequency of the white noise. A large oscillation followed by a steady state offset is observed when the system is submitted to a ramp like variation in density. The steady state offset is a characteristic response seen when ramp like disturbances are applied to systems with a MPC controller with an integrated white noise models [17]. A step change in density causes a large oscillation on the controlled variables, but this effect is compensated by the control system after approximately 100 s. Note that, a relatively high magnitude of disturbance has been introduced to evaluate the capability of the controller under extreme scenario. In practice, the magnitude of the disturbance will be relatively less and therefore the controller performance will be better.

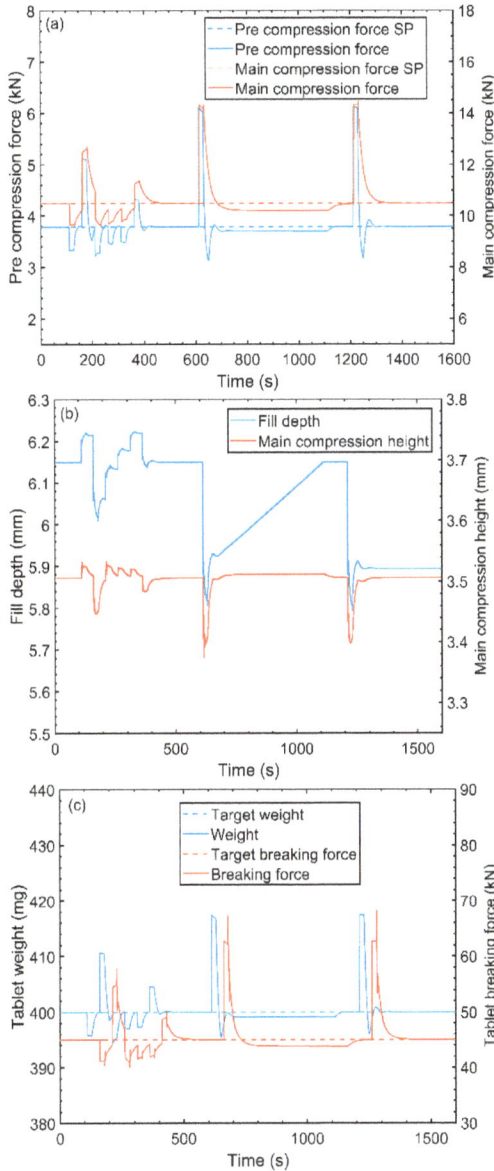

Figure 12. Control strategy 1—Disturbance rejection scenario. (**a**) Critical process parameters (**b**) Actuators (**c**) Critical quality attributes.

The main advantage of this strategy is the fact that it does not require the use of online real-time sensors for tablet weight and breaking force, since these critical quality attributes are controlled indirectly through the compression forces. Another advantage of this strategy is its relatively fast response time when compared to strategies that involve a cascade arrangement of controllers. A major disadvantage of this control strategy is the fact that it requires a relationship between the compression force and tablet CQAs to be established experimentally during the setup of the system. In order to avoid tablet defects such as capping, it is necessary to tune the controller at the adequate pre-compression force and tablet weight ranges. The relation between the pre-compression force and tablet weight ranges can be manipulated by changing the pre-compression height at which the tablet press operates. It is important to note that the controller model should be regenerated whenever changes in pre-compression height are made. A few specific disturbances such as fluctuations in lubricant concentration may also cause the relation between main compression force and tablet breaking force to shift, leading to fluctuations in tablet breaking force which cannot be mitigated by this control strategy. The best way to avoid these fluctuations is by ensuring control of the blend composition, as proposed by Singh et al. [30].

This control strategy has also been implemented in the pilot-plant for comparison with the simulation results. The effect that changes in the compression force set points have on tablet weight and breaking force was studied using only the simulation tool, since no experimental dynamic data could be obtained for those variables. The comparison is shown in Figure 13. Experimental results match the simulations, demonstrating the capabilities of the developed model for control system design.

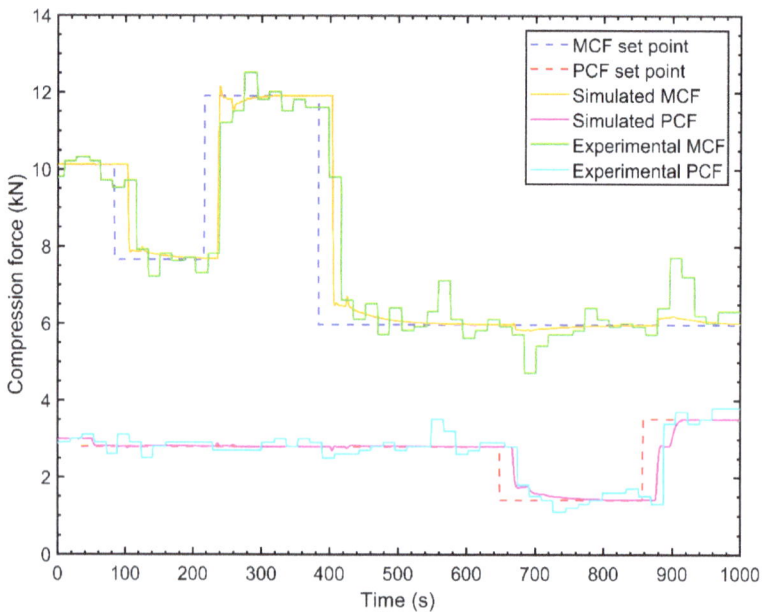

Figure 13. Control strategy 1—Experimental validation.

7.2.2. Strategy 2—Cascade Control of Weight and Breaking Force

The dynamic response of the simulated system under a set point tracking scenario is shown in Figure 14. Both tablet weight and tablet breaking force are able to successfully track their set points by actuating on the compression forces. Oscillations are observed in the first and second tablet weight set point changes. These oscillations are also present in pre-compression force and can be explained by a model mismatch in the pre-compression force model due to the non-linear behavior of this variable, which becomes more pronounced as it moves away from the operating point at which the controller was tuned. Spikes in tablet breaking force are seen in Figure 14a as a response to changes in the CQAs set points. These spikes can be avoided by limiting the maximum rate of change of the manipulated variables of the slave controller.

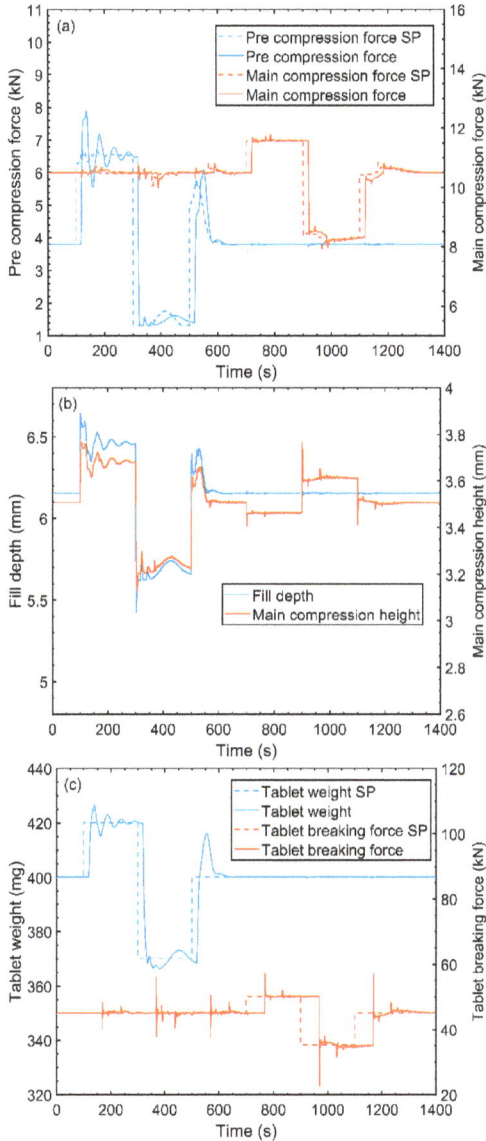

Figure 14. Control strategy 2—Set point tracking scenario. (**a**) Critical process parameters (**b**) Actuators (**c**) Critical quality attributes. SP: Set point.

The system's reaction to a disturbance rejection scenario with changes in density is similar to what was observed for control strategy 1 and is shown in Figure 15. The main difference between the two strategies evident when a ramp disturbance is applied to density. Differently than what is seen in strategy 1, no offset between the CQAs set points and actual values are observed in strategy 2. This occurs because integrator of the inner control loop partially absorbs the ramp perturbation, yielding a step like disturbance in its controlled variables. The step disturbance is then completely eliminated by the master control loop, causing no offset to be observed between the CQAs set points and actual values.

Control strategy 2 advantages lie in the fact that tablet weight and breaking force are monitored in real time and controlled through a cascade arrangement. This arrangement allows the usage of tablet weight and breaking force sensors with relatively low sampling rate. Directly sensing of tablet CQAs eliminates the need of sometimes inaccurate inferences of these parameters from the compression forces. Certain disturbances are also more adequately handled by this control architecture as described above. The main disadvantages of this control strategy are that it adds another element of non-linearity to the system by relating tablet weight to pre-compression force, which is non-linearly related to its actuator (fill depth). These accumulated non-linearities result in a narrower stability margin. Control strategy 2, due to its cascade arrangement, has slightly longer response times when compared to strategy 1, which can be considered a minor disadvantage. It is important to note that, just as in strategy 1, it is necessary to tune the compression force controller with the adequate pre-compression force range in mind.

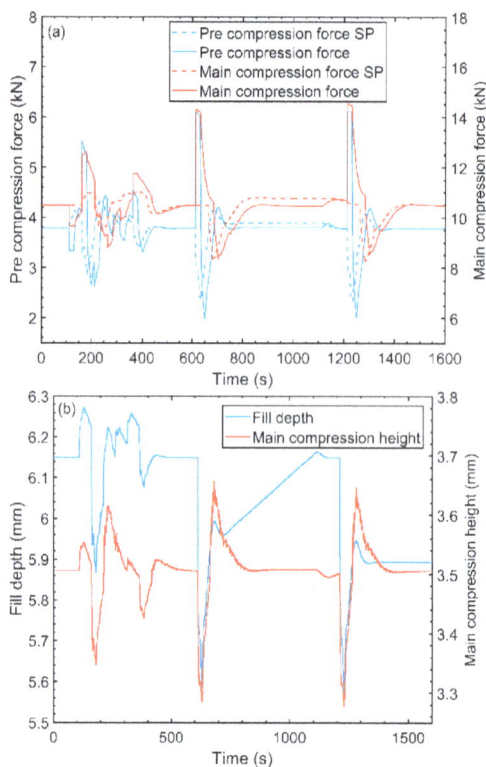

Figure 15. *Cont.*

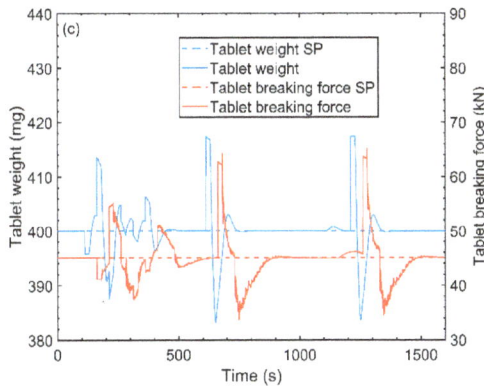

Figure 15. Control strategy 2—Disturbance rejection scenario. (**a**) Critical process parameters (**b**) Actuators (**c**) Critical quality attributes.

7.2.3. Strategy 3—Direct Control of Tablet Weight and Cascade Tablet Breaking Force Control

Figure 16 shows the systems response in a set point tracking scenario. Similar step responses in tablet weight are seen in all three weight set point changes, which is a characteristic of a linear system. An interaction between tablet weight and breaking force can be noticed in Figure 16a in the form of peaks. Changes in tablet weight set point result in fill depth manipulations, these manipulations lead to change in main compression height in order to maintain the main compression ratio at the desired value. There is a slight offset between the dynamics of fill depth and main compression height in the tablet press model, which causes the peaks seen in Figure 16a. Tablet breaking force is able to track its set points when no changes in tablet weight are made.

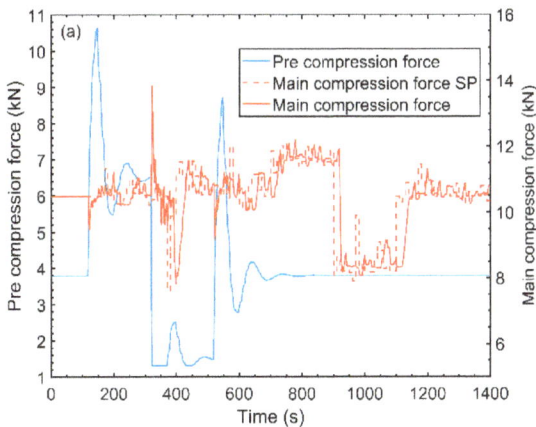

Figure 16. *Cont.*

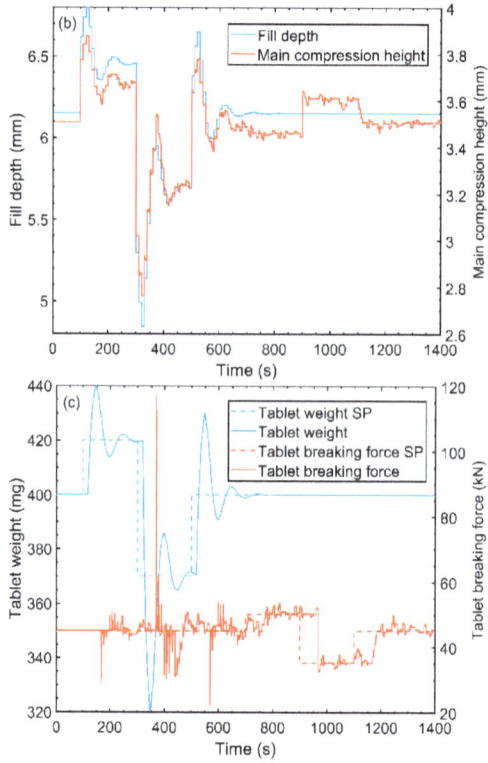

Figure 16. Control strategy 3—Set point tracking scenario. (**a**) Critical process parameters (**b**) Actuators (**c**) Critical quality attributes.

The system's response to disturbances in powder density is shown in Figure 17. The response is similar to what was observed for strategy 1.

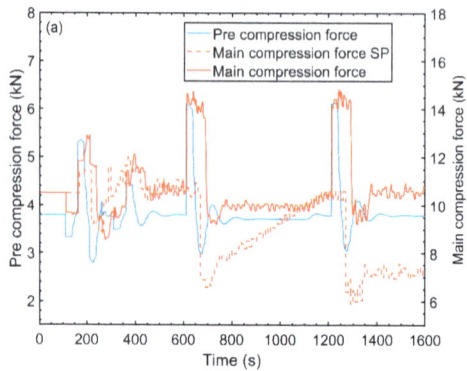

Figure 17. *Cont.*

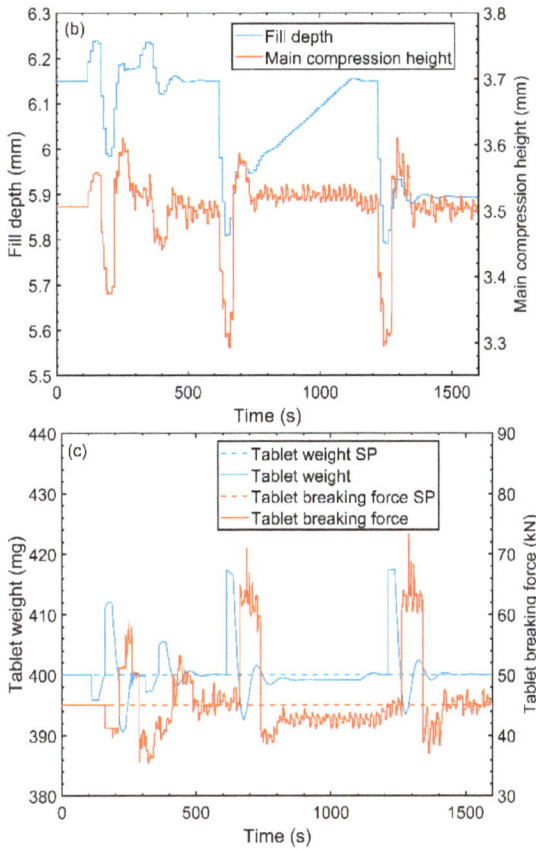

Figure 17. Control strategy 3—Disturbance rejection scenario. (**a**) Critical process parameters (**b**) Actuators (**c**) Critical quality attributes.

The main advantage of control strategy 3 lies in the fact that the inner main compression controller isolates the non-linear behavior of the compression forces from the master controller. The master controller actuates on variables that linearly affect tablet weight and breaking force. In this strategy, pre-compression force can be varied or controlled independently without affecting any tablet CQAs, making it easier to avoid tablet defects related to inadequate dwell times. Theoretically, if the controller models where adequately regressed, control strategy 3 should have the best performance among the three evaluated strategies. This strategy can be further improved by using fill depth as a measured disturbance signal in the main compression controller.

8. Discussion and Conclusions

A control relevant model for the tablet compaction unit was developed and validated. The developed model was able to successfully capture the non-linear behavior of the compaction process as well as tablet production rate dependent transport delays. The relevance of this model lies in its fast execution time while still accurately representing the dynamic behavior of the process.

The applicability of the model for control system design was evaluated and the developed control strategies were implemented on an experimental setup. A comparison between three control algorithms was conducted leading to the conclusion that an MPC with an unmeasured disturbance model was the most adequate control algorithm for the studied system. Three different strategies were explored. Their characteristics as well as their possible applications were studied through a set point tracking scenario and a disturbance rejection scenario. It was concluded that the selection of control strategies for a given compaction process is heavily depended on the availability and quality of real time data of tablet critical quality attributes. As far as the authors' knowledge goes, this is the first work that involved the development of a control strategy for a tablet compaction unit in a simulation environment follow by experimental implementation.

This work should serve as a major stepping-stone for establishment of a fully integrated and controlled direct compaction process in commercial scale, leading to the achievement of true Quality by Control. The future work includes the development and implementation of RTD (residence time distribution) based control system.

Acknowledgments: This work is supported by the Rutgers Research Council, through grant 202342 RC-17-Singh R, the US Food and Drug Administration (FDA), through grant 5U01FD005535 and National Science Foundation Engineering Research Center on Structured Organic Particulate Systems, through Grant NSF-ECC 0540855.

Author Contributions: All the authors contributed equally to this work.

Conflicts of Interest: The authors declare no conflict of interest. The founding sponsors had no role in the design of the study; in the collection, analyses, or interpretation of data; in the writing of the manuscript and in the decision to publish the results.

References

1. U.S. Department of Health and Human Services Food and Drug Administration Guidance for Industry: ICH Q8(R2) Pharmaceutical Development. *Work. Qual. Des. Pharm.* **2009**, *8*, 28.
2. International conference on harmonisation of technical requirements for registration of pharmaceuticals for human use Pharmaceutical Development Q8(R2). In Proceedings of the ICH Harmonised Tripartite Guideline, Kuala Lumpur, Malaysia, 26–28 July 2010; pp. 1–28.
3. Mesbah, A.; Paulson, J.A.; Lakerveld, R.; Braatz, R.D. Model Predictive Control of an Integrated Continuous Pharmaceutical Manufacturing Pilot Plant. *Org. Process Res. Dev.* **2017**, *21*, 844–854. [CrossRef]
4. Sen, M.; Chaudhury, A.; Singh, R.; John, J.; Ramachandran, R. Multi-scale flowsheet simulation of an integrated continuous purification-downstream pharmaceutical manufacturing process. *Int. J. Pharm.* **2013**, *445*, 29–38. [CrossRef] [PubMed]
5. Boukouvala, F.; Niotis, V.; Ramachandran, R.; Muzzio, F.J.; Ierapetritou, M.G. An integrated approach for dynamic flowsheet modeling and sensitivity analysis of a continuous tablet manufacturing process. *Comput. Chem. Eng.* **2012**, *42*, 30–47. [CrossRef]
6. Jain, S. Mechanical properties of powders for compaction and tableting: An overview. *Pharm. Sci. Technol. Today* **1999**, *2*, 20–31. [CrossRef]
7. Michaut, F.; Busignies, V.; Fouquereau, C.; Huet de Barochez, E.; Leclerc, B.; Tchoreloff, P. Evaluation of a Rotary Tablet Press Simulator as a Tool for the Characterization of Compaction Properties of Pharmaceutical Products. *J. Pharm. Sci.* **2010**, *99*, 2874–2885. [CrossRef] [PubMed]
8. Sinka, I.C.; Cunningham, J.C.; Zavaliangos, A. The effect of wall friction in the compaction of pharmaceutical tablets with curved faces: A validation study of the Drucker-Prager Cap model. *Powder Technol.* **2003**, *133*, 33–43. [CrossRef]
9. Mendez, R.; Muzzio, F.; Velazquez, C. Study of the effects of feed frames on powder blend properties during the filling of tablet press dies. *Powder Technol.* **2010**, *200*, 105–116. [CrossRef]
10. Akande, O.F.; Rubinstein, M.H.; Ford, J.L. Examination of the compaction properties of a 1:1 acetaminophen: Microcrystalline cellulose mixture using precompression and main compression. *J. Pharm. Sci.* **1997**, *86*, 900–907. [CrossRef] [PubMed]
11. Gonnissen, Y.; Gonçalves, S.I.V.; De Geest, B.G.; Remon, J.P.; Vervaet, C. Process design applied to optimise a directly compressible powder produced via a continuous manufacturing process. *Eur. J. Pharm. Biopharm.* **2008**, *68*, 760–770. [CrossRef] [PubMed]
12. Haware, R.V.; Tho, I.; Bauer-Brandl, A. Application of multivariate methods to compression behavior evaluation of directly compressible materials. *Eur. J. Pharm. Biopharm.* **2009**, *72*, 148–155. [CrossRef] [PubMed]

13. Seitz, J.A.; Flessland, G.M. Evaluation of the Physical Properties of Compressed Tablets I: Tablet Hardness and Friability. *J. Pharm. Sci.* **1965**, *54*, 1353–1357. [CrossRef] [PubMed]

14. Mayne, D.Q. Model predictive control: Recent developments and future promise. *Automatica* **2014**, *50*, 2967–2986. [CrossRef]

15. Di Cairano, S. An Industry perspective on MPC in large volumes applications: Potential benefits and open challenges. *IFAC Proc. Vol.* **2012**, *45*, 52–59. [CrossRef]

16. Qin, S.J.; Badgwell, T.A. A survey of industrial model predictive control technology. *Control Eng. Pract.* **2003**, *11*, 733–764. [CrossRef]

17. Zagrobelny, M.A. MPC Performance Monitoring and Disturbance Model Identification, University of Winsconsin-Madison. *Diss. Theses Gradworks* **2014**. Available online: http://jbrwww.che.wisc.edu/theses/zagrobelny.pdf (accessed 30 November 2017).

18. Mayne, D.Q.; Raković, S.V.; Findeisen, R.; Allgöwer, F. Robust output feedback model predictive control of constrained linear systems. *Automatica* **2006**, *42*, 1217–1222. [CrossRef]

19. Mayne, D.Q.; Rawlings, J.B.; Rao, C.V.; Scokaert, P.O.M. Constrained model predictive control: Stability and optimality. *Automatica* **2000**, *36*, 789–814. [CrossRef]

20. Streif, S.; Kögel, M.; Bäthge, T.; Findeisen, R. Robust Nonlinear Model Predictive Control with Constraint Satisfaction: A Relaxation-based Approach. *FAC World Congr.* **2014**, *47*, 11073–11079. [CrossRef]

21. Maiworm, M.; Bäthge, T.; Findeisen, R. Scenario-based Model Predictive Control: Recursive feasibility and stability. *IFAC PapersOnLine* **2015**, *28*, 50–56. [CrossRef]

22. Allgöwer, F.; Findeisen, R.; Ebenbauer, C. Nonlinear model predictive control. In *Control Systems, Robotics and Automation*; Birkhäuser: Basel, Basel-Stadt, Switzerland, 2009; Volume XI.

23. Su, Q.; Moreno, M.; Giridhar, A.; Reklaitis, G.V.; Nagy, Z.K. A Systematic Framework for Process Control Design and Risk Analysis in Continuous Pharmaceutical Solid-Dosage Manufacturing. *J. Pharm. Innov.* **2017**, *12*, 1–20. [CrossRef]

24. Singh, R.; Ierapetritou, M.; Ramachandran, R. System-wide hybrid MPC–PID control of a continuous pharmaceutical tablet manufacturing process via direct compaction. *Eur. J. Pharm. Biopharm.* **2013**, *85*, 1164–1182. [CrossRef] [PubMed]

25. Singh, R.; Ierapetritou, M.; Ramachandran, R. An engineering study on the enhanced control and operation of continuous manufacturing of pharmaceutical tablets via roller compaction. *Int. J. Pharm.* **2012**. [CrossRef] [PubMed]

26. Rehrl, J.; Kruisz, J.; Sacher, S.; Khinast, J.; Horn, M. Optimized continuous pharmaceutical manufacturing via model-predictive control. *Int. J. Pharm.* **2016**, *510*, 100–115. [CrossRef] [PubMed]

27. Zhao, X.J.; Gatumel, C.; Dirion, J.L.; Berthiaux, H.; Cabassud, M. Implementation of a control loop for a continuous powder mixing process. In Proceedings of the AIChE 2013 Annual Meeting, San Francisco, CA, USA, 6 November 2013.

28. Singh, R.; Román-Ospino, A.D.; Romañach, R.J.; Ierapetritou, M.; Ramachandran, R. Real time monitoring of powder blend bulk density for coupled feed-forward/feed-back control of a continuous direct compaction tablet manufacturing process. *Int. J. Pharm.* **2015**, *495*, 612–625. [CrossRef] [PubMed]

29. Singh, R.; Muzzio, F.; Ierapetritou, M.; Ramachandran, R. A Combined Feed-Forward/Feed-Back Control System for a QbD-Based Continuous Tablet Manufacturing Process. *Processes* **2015**, *3*, 339–356. [CrossRef]

30. Singh, R.; Sahay, A.; Karry, K.M.; Muzzio, F.; Ierapetritou, M.; Ramachandran, R. Implementation of an advanced hybrid MPC-PID control system using PAT tools into a direct compaction continuous pharmaceutical tablet manufacturing pilot plant. *Int. J. Pharm.* **2014**, *473*, 38–54. [CrossRef] [PubMed]

31. Sen, M.; Singh, R.; Ramachandran, R. A Hybrid MPC-PID Control System Design for the Continuous Purification and Processing of Active Pharmaceutical Ingredients. *Processes* **2014**, *2*, 392–418. [CrossRef]

32. Bhaskar, A.; Barros, F.N.; Singh, R. Development and implementation of an advanced model predictive control system into continuous pharmaceutical tablet compaction process. *Int. J. Pharm.* **2017**, *534*, 159–178. [CrossRef] [PubMed]

33. Singh, R.; Sahay, A.; Muzzio, F.; Ierapetritou, M.; Ramachandran, R. A systematic framework for onsite design and implementation of a control system in a continuous tablet manufacturing process. *Comput. Chem. Eng.* **2014**, *66*, 186–200. [CrossRef]

34. Sarkar, S.; Ooi, S.M.; Liew, C.V.; Heng, P.W.S. Influence of Rate of Force Application during Compression on Tablet Capping. *J. Pharm. Sci.* **2015**, *104*, 1319–1327. [CrossRef] [PubMed]

35. Seborg, D.; Edgar, T.; Mellichamp, D. *Process Dynamics and Control*, 2nd ed.; Wiley: Hoboken, NJ, USA, 2004; ISBN 0-471-00077-9.

36. Patel, S.; Kaushal, A.M.; Bansal, A.K. Compression Physics in the Formulation Development of Tablets. *Crit. Rev. Ther. Drug Carr. Syst.* **2006**, *23*, 1–66. [CrossRef]

37. Kawakita, K.; Lüdde, K.H. Some considerations on powder compression equations. *Powder Technol.* **1971**, *4*, 61–68. [CrossRef]

Article

Economic Benefit from Progressive Integration of Scheduling and Control for Continuous Chemical Processes

Logan D. R. Beal [1], Damon Petersen [1], Guilherme Pila [1], Brady Davis [1], Sean Warnick [2] and John D. Hedengren [1],*

[1]   Department of Chemical Engineering, Brigham Young University, Provo, UT 84602, USA;
      beal.logan@gmail.com (L.D.R.B.); damon.chem.e@gmail.com (D.P.); guipila@gmail.com (G.P.);
      bradyrdavis@gmail.com (B.D.)
[2]   Department of Computer Science, Brigham Young University, Provo, UT 84602, USA;
      sean.warnick@gmail.com
*   Correspondence: john.hedengren@byu.edu

Received: 14 November 2017; Accepted: 7 December 2017; Published: 13 December 2017

Abstract: Performance of integrated production scheduling and advanced process control with disturbances is summarized and reviewed with four progressive stages of scheduling and control integration and responsiveness to disturbances: open-loop segregated scheduling and control, closed-loop segregated scheduling and control, open-loop scheduling with consideration of process dynamics, and closed-loop integrated scheduling and control responsive to process disturbances and market fluctuations. Progressive economic benefit from dynamic rescheduling and integrating scheduling and control is shown on a continuously stirred tank reactor (CSTR) benchmark application in closed-loop simulations over 24 h. A fixed horizon integrated scheduling and control formulation for multi-product, continuous chemical processes is utilized, in which nonlinear model predictive control (NMPC) and continuous-time scheduling are combined.

Keywords: scheduling; model predictive control; dynamic market; market fluctuations; process disturbances; nonlinear; integration

1. Introduction

Production scheduling and advanced process control are related tasks for optimizing chemical process operation. Traditionally, implementation of process control and scheduling are separated; however, research suggests that opportunity is lost from separate implementation [1–3]. Many researchers suggest that economic benefit may arise from integrating production scheduling and process control [4–10]. Though integration may provide economic benefit, scheduling and control integration presents several challenges which are outlined in multiple reviews on integrated scheduling and control (ISC) [3,11–14]. Some of the major challenges to integration mentioned in review articles include time-scale bridging, computational burden, and human factors such as organizational and behavioral challenges.

1.1. Economic Benefit from Integrated Scheduling and Control

Many complex, interrelated elements factor into the potential benefit from the integration of scheduling and control, including the following [3,11]:

(i)   Rapid fluctuations in dynamic product demand;
(ii)   Rapid fluctuations in dynamic energy rates;
(iii) Dynamic production costs;

(iv) Benefits of increased energy efficiency;

(v) Necessity of control-level dynamics information for optimal production schedule calculation.

In the current economic environment, demand and selling prices for the products and inputs of chemical processes can change significantly over the course of not only months and years, but on the scales of weeks, days, and hours [3,11,12]. Energy rates often fluctuate hourly, with peak pricing during peak demand hours and rate cuts during off peak hours (sometimes even negative rate cuts occur during periods of excess energy production) [11]. An optimal schedule is intrinsically dependent upon market conditions such as input material price, product demand and pricing, and energy rates [12]. Therefore, when market conditions change, the optimal production sequence or schedule may also change. Since the time scale at which market factors fluctuates has decreased, the time scale at which scheduling decisions must be recalculated should also decrease [3,12].

Frequent recalculation of scheduling on a time scale closer to that of advanced process control (seconds to minutes) leads to a greater need to integrate process dynamics into the scheduling problem [3]. According to a previous review [11], process dynamics are important for optimal production scheduling because (i) transition times between any given products are determined by process dynamics and process control; (ii) process dynamics may show that a calculated production sequence or schedule is operationally infeasible; and (iii) process disturbances may cause a change in the optimal production sequence or schedule.

1.2. Previous Work

Significant research has been conducted on the integration of production scheduling and advanced process control [3,11]. This section summarizes evidence for economic benefit from integration, upon which this work builds. Previous research showing the benefits of combined scheduling and control is explored and previous research done to show the economic benefits of combined over segregated scheduling and control is examined. The reviewed articles are summarized in Table 1. This work focuses on research demonstrating benefit over a baseline comparison of segregated scheduling and control (SSC).

Table 1. Economic benefit of integrated scheduling and control (ISC) over segregated scheduling and control (SSC) (CSTR: continuously stirred tank reactor; MMA: methyl methacrylate; DR: demand response; FRB: fluidized bed reactor; RTN: resource task network; ASU: air separation unit; HIPS: high-impact polystyrene; PFR: plug flow reactor; SISO: single-input single-output; MIMO: multiple-input multiple-output).

Author	Shows Benefit of ISC over SSC	Batch Process	Continuous Process	Example Application (s)
Baldea et al. (2015) [15]			X	CSTR
Baldea et al. (2016) [16]			X	MMA
Baldea (2017) [17]			X	DR chemical processes and power generation facilities
Beal (2017) [18]			X	CSTR
Beal (2017) [19]			X	CSTR
Beal (2017a) [20]			X	CSTR
Cai et al. (2012) [21]		X		Semiconductor production
Capon-Garcia et al. (2013) [6]		X		2 different batch plants (1-stage, 3-product & 3-stage, 3-product)
Chatzidoukas et al. (2003) [22]	X		X	gas-phase polyolefin FBR.
Chatzidoukas et al. (2009) [23]	X		X	catalytic olefin copolymerization FBR
Chu & You (2012) [24]			X	MMA
Chu & You (2013) [25]			X	CSTR
Chu & You (2013a) [26]		X		polymerization with parallel reactors & 1 purification unit (RTN)
Chu & You (2013b) [27]	X	X		5-unit batch process
Chu & You (2013c) [28]	X	X		sequential batch process
Chu & You (2014) [29]		X		batch process (reaction task, filtration task, reaction task)

Table 1. *Cont.*

Author	Shows Benefit of ISC over SSC	Batch Process	Continuous Process	Example Application (s)
Chu & You (2014a) [30]		X		8-unit batch process
Chu & You (2014b) [31]	X	X		8-unit batch process
Dias et al. (2016) [32]			X	MMA
Du et al. (2015) [33]			X	CSTR & MMA
Flores-Tlacuahuac & Grossmann (2006) [34]			X	CSTR
Flores-Tlacuahuac (2010) [8]			X	Parallel CSTRs
Gutiérrez-Limón et al. (2011) [35]			X	CSTR
Gutiérrez-Limón et al. (2016) [36]			X	CSTR & MMA
Gutiérrez-Limón & Flores-Tlacuahuac (2014) [37]			X	CSTR
Koller & Ricardez-Sandoval (2017) [38]			X	CSTR
Nie & Bieglier (2012) [7]	X	X		flowshop plant (batch reactor, filter, distillation column)
Nie et al. (2015) [39]		X	X	polymerization with parallel reactors & 1 purification unit
Nystrom et al. (2005) [40]			X	industrial polymerization process
Nystrom et al. (2006) [4]			X	industrial polymerization process
Patil et al. (2015) [41]			X	CSTR & HIPS
Pattison et al. (2016) [42]	X		X	ASU model
Pattison et al. (2017) [10]			X	ASU model
Prata (2008) et al. [43]			X	medium industry-scale model
Terrazas-Moreno et al. (2008) [44]			X	MMA (with one CSTR) & HIPS
Terrazas-Moreno & Flores-Tlacuahuac (2007) [45]			X	HIPS & MMA
Terrazas-Moreno & Flores-Tlacuahuac (2008) [9]			X	HIPS & MMA
You & Grossmann (2008) [46]			X	medium and large polystyrene supply chaiins
Zhuge & Ierapetritou (2012) [47]			X	CSTR & PFR.
Zhuge & Ierapetritou (2014) [48]		X		simple and complex batch processes
Zhuge & Ierapetritou (2015) [49]			X	SISO & MIMO CSTRs
Zhuge & Ierapetritou (2016) [50]	X		X	CSTR & MMA

1.2.1. Integrating Process Dynamics into Scheduling

Mahadevan et al. suggest that process dynamics should be considered in scheduling problems. To avoid the computational requirements of mixed-integer nonlinear programming (MINLP), they include process dynamics as costs in the scheduling problem [51]. Flores-Tlacuahuac and Grossman implement process dynamics into scheduling directly in a mixed-integer dynamic optimization (MIDO) problem with a continuous stirred tank reactor (CSTR). Chatzidoukas et al. demonstrated the economic benefit of implementing scheduling in a MIDO problem for polymerization, solving product grade transitions along with the scheduling problem [22]. Economic benefit has also been shown for simultaneous selection of linear controllers for grade transitions and scheduling, ensuring that the process dynamics from the controller selection are accounted for in the scheduling problem [23]. Terrazas-Moreno et al. also demonstrate the benefits of process dynamics in cyclic scheduling for continuous chemical processes [45]. Capon-Garcia et al. prove the benefit of implementing process dynamics in batch scheduling via an MIDO problem [6]. MIDO batch scheduling optimization with dynamic process models is shown to be more profitable than a fixed-recipe approach. Chu and You also demonstrate enhanced performance from batch scheduling with simultaneous solution of dynamic process models over a traditional batch scheduling approach [27,28,31]. Economic benefit from integrating process dynamics into batch and semi-batch scheduling has also been demonstrated via mixed-logic dynamic optimization in state equipment networks and solution with Benders decomposition in resource task networks [7,39]. Potential for economic benefit from integrating process dynamics into design, scheduling, and control problems has also been demonstrated [41,44,52]. Computational reduction of incorporating process dynamics into scheduling has been investigated successfully, maintaining benefit from the incorporation of process dynamics into scheduling while reducing dynamic model order [10,15,16,33,42].

1.2.2. Reactive Integrated Scheduling and Control

Research indicates that additional benefit arises from ISC responsive to process disturbances, which are a form of process uncertainty. This is in congruence with recent work by Gupta and Maravelias demonstrating that increased frequency of schedule rescheduling (online scheduling) can improve process economics [53–55]. Many previous works considering reactive ISC are outlined in Table 2. For a complete review of ISC under uncertainty, the reader is directed to a recent review by Dias and Ierapetritou [32]. Zhuge and Ierapetritou demonstrate increased profit from closed-loop implementation (over open-loop implementation) of combined scheduling and control in the presence of process disturbances [47]. The schedule is optimally recalculated when a disturbance is encountered. Zhuge and Ierapetritou also present methodology to reduce the computational burden of ISC to enable closed-loop online operation for batch and continuous processes. They propose using multi-parametric model predictive control for online batch scheduling and control [48], fast model predictive control coupled with reduced order (piece-wise affine) models in scheduling and control for continuous processes [49], and decomposition into separate problems for continuous processes [50]. Chu and You demonstrate the economic benefit of closed-loop moving horizon scheduling with consideration of process dynamics in batch scheduling [29]. Chu and You also investigate the reduction of computational burden to enable online closed-loop ISC for batch and continuous processes. They investigate utilization of Pareto frontiers to decompose batch scheduling into an online mixed-integer linear programming (MILP) problem and offline dynamic optimization (DO) problems [26]. Investigation of a solution via mixed-integer nonlinear fractional programming and Dinkelbach's algorithm coupled with decomposing into an online scheduling and controller selection and offline transition time calculation [24].

Table 2. Works considering reactive ISC.

Authors	Product Price Disturbance	Product Demand Disturbance	Process Variable Disturbance	Other Disturbances
Baldea et al. (2016) [16]	X	X		
Baldea (2017) [17]			X	
Cai et al. (2012) [21]			X	
Chu & You (2012) [24]			X	
Du et al. (2015) [33]				
Flores-Tlacuahuac (2010) [8]		X		
Gutiérrez-Limón et al. (2016) [36]		X		
Kopanos & Pistikopoulos (2014) [56]			X	
Liu et al. (2012) [57]	X	X		
Patil et al. (2015) [41]			X	
Pattison et al. (2017) [10]		X	X	
Touretzky & Baldea (2014) [58]				Weather & energy price
You & Grossmann (2008) [46]		X		
Zhuge & Ierapetritou (2012) [47]			X	
Zhuge & Ierapetritou (2015) [49]			X	

Closed-loop reactive ISC responds to process uncertainty in a reactive rather than preventative manner [59]. Preventative approaches to dealing with process uncertainty in ISC have also been investigated. Chu and You investigated accounting for process uncertainty in batch processes in a two-stage stochastic programming problem solved by a generalized Benders decomposition [28]. The computational requirements of the problem prevent online implementation. Dias and Ierapetritou demonstrate the benefits of using robust model predictive control in ISC to optimally address process uncertainty in continuous chemical processes [32].

1.2.3. Responsiveness to Market Fluctuations

As mentioned in Section 1.1, a major consideration affecting the profitability of ISC is rapidly fluctuating market conditions. If the market changes, the schedule should be reoptimized to new market demands and price forecasts. This is again congruent with recent work demonstrating benefit from frequent re-scheduling [53–55]. Literature on ISC reactive to market fluctuations is relatively limited in scope. Gutierrez-Limon et al. demonstrated integrated planning, scheduling, and control responsive to fluctuations in market demand on a CSTR benchmark application [36,37]. Pattison et al. investigated ISC with an air separation unit (ASU) in fast-changing electricity markets, responding optimally to price fluctuations [42]. Pattison et al. also demonstrated theoretical developments with moving horizon closed-loop scheduling in volatile market conditions [10]. Periodic rescheduling to account for fluctuating market conditions was implemented successfully on an ASU application.

1.3. Purpose of This Work

This work aims to provide evidence for the progressive economic benefits of combining scheduling and control and operating combined scheduling and control in a closed-loop responsive to disturbances over segregated scheduling and control and open-loop formulations for continuous chemical processes. This work demonstrates the benefits of integration through presenting four progressive stages of integration and responsiveness to disturbances. This work comprehensively demonstrates the progression of economic benefit from (1) integrating process dynamics and control level information into production scheduling and (2) closed-loop integrated scheduling and control responsive to market fluctuations. Such a comprehensive examination of economic benefit has not been performed to the authors' knowledge. This work also utilizes a novel, computationally light decomposed integration method employing continuous-time scheduling and nonlinear model predictive control (NMPC) as the fourth phase of integration. This method is outlined in detail in another work [60]. Although the phases of integration presented in this work are not comprehensively representative of integration methods presented in the literature, the concepts of integration progressively applied in the four phases are applicable across the majority of formulations in the literature.

2. Phases of Progressive Integration

This section introduces the four phases of progressive integration of scheduling and control investigated in this work. Each phase is outlined in the appropriate section.

2.1. Phase 1: Fully Segregated Scheduling and Control

A schedule is created infrequently (every 24 h in this work) and a controller seeks to implement the schedule throughout the 24 h with no other considerations. In this format, the schedule is open-loop, whereas the control is closed-loop. The controller acts to reject disturbances and process noise to direct the process to follow the predetermined schedule (see Figure 1).

This work considers an NMPC controller and a continuous-time, slot-based schedule (Section 2.5). For this phase, the schedule is uninformed of transition times as dictated by process dynamics and control structure. All product grade transitions are considered to produce a fixed amount of off-specification material and to require the same duration.

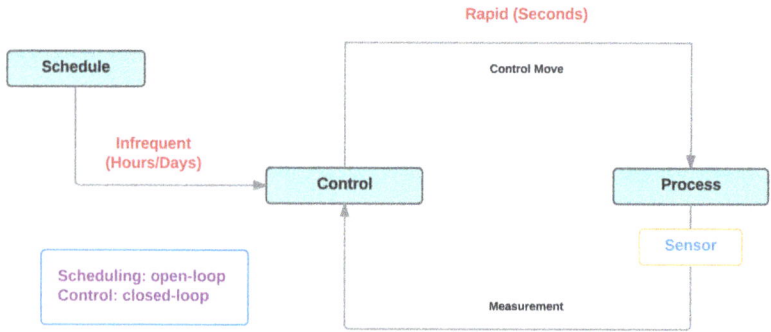

Figure 1. Phase 1: Open-loop scheduling determined once per day with no consideration of process dynamics. Closed-loop control implemented to follow the schedule.

2.2. Phase 2: Reactive Closed-Loop Segregated Scheduling and Control

Phase two is a closed-loop implementation of completely segregated scheduling and control. The formulation for Phase 2 is identical to that of Phase 1 with the exception that the schedule is recalculated in the event of a process disturbance or market update (see Figure 2).

2.3. Phase 3: Open-Loop Integrated Scheduling and Control

For phase 3, the schedule is calculated infrequently, similar to phase 1 (every 24 h in this work). However, information about the control structure and process dynamics in the form of transition times are fed to the scheduling algorithm to enable a more intelligent decision. Scheduling remains open-loop while the controller remains closed-loop to respond to noise and process disturbances while implementing the schedule (see Figure 3).

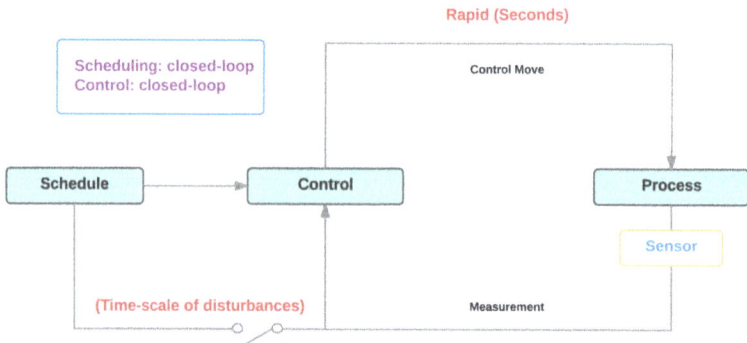

Figure 2. Phase 2: Dual-loop segregated scheduling and control. Scheduling is recalculated reactively in the presence of process disturbances above a threshold or updated market conditions. Closed-loop control implements the schedule in the absence of disturbances.

This work considers a continuous-time schedule with process dynamics incorporated via transition times estimated by NMPC. Transitions between products are simulated with a dynamic process model and nonlinear model predictive controller implementation. The time required to transition between products is minimized by the controller, and the simulated time required to transition is fed to the scheduler as an input to the continuous-time scheduling formulation (Section 2.5).

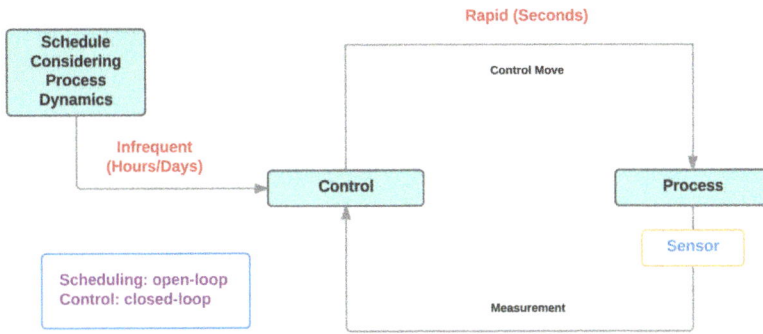

Figure 3. Phase 3: Open-loop scheduling determined once per day with consideration of process dynamics and control structure in the form of grade transition information. Closed-loop control implemented to follow the schedule.

2.4. Phase 4: Closed-Loop Integrated Scheduling and Control Responsive to Market Fluctuations

Phase 4 represents closed-loop implementation of ISC responsive to both market fluctuations and process disturbances. This work utilizes the formulation for computationally light online scheduling and control for closed-loop implementation introduced in another work by the authors [60]. As in phase 3, a continuous-time schedule is implemented with NMPC-estimated transition times as inputs to the scheduling optimization; however, the ISC algorithm is implemented not only once at the beginning of the horizon as in phase 3, but triggered by updated market conditions or process disturbances above a threshold (see Figure 4). This enables the ISC algorithm to respond to fluctuations in market conditions as well as respond to measured process disturbances in a timely manner to ensure that production scheduling and control are updated to reflect optimal operation with current market conditions and process state.

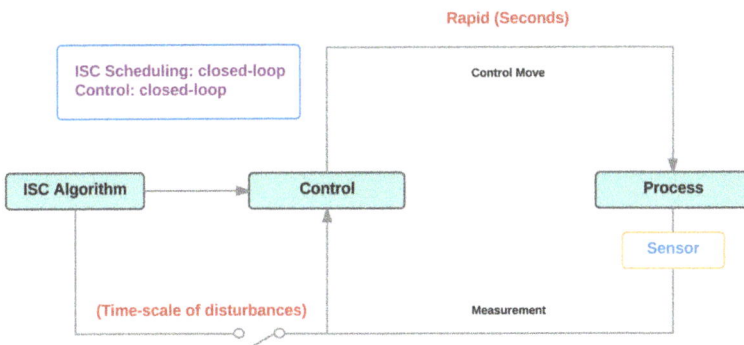

Figure 4. Phase 4: Closed-loop combined scheduling and control responsive to both process disturbances and updated market information.

The formulation for phase 4 builds on the work of Zhuge et al. [49], which justifies decomposing slot-based ISC into two subproblems: (1) NLP solution of transition times and transition control profiles and (2) MILP solution of the slot-based, continuous-time schedule. The formulation in [60] expands the work of Zhuge et al. by combining a look-up transition time table with control profiles and transition times between known product steady-state conditions, calculated offline and stored in memory, with transitions from current conditions to each product. The transitions from current conditions or most recently received process measurements are the only transition times and transition control profiles required to be solved at each iteration of combined scheduling and control (Equations (2)–(8)). This reduces the online problem to few nonlinear programming (NLP) dynamic optimization problems and an MILP problem only, eliminating the computational requirements of MINLP. This work also introduces the use of nonlinear models in this form of decomposition. Zhuge et al. use piecewise affine (PWA) models, whereas this work harnesses full nonlinear process dynamics to calculate optimal control and scheduling.

This work also builds on the work of Pattison et al., who demonstrate closed-loop moving horizon combined scheduling and control to respond to market updates [10]. This formulation, however, does not use simplified dynamic process models for scheduling, but rather maintains nonlinear process dynamics while reducing computational burden via problem decomposition into offline and online components and further decomposition of the problem into computationally light NLP and MILP problems, solvable together without the need for iterative alternation [60].

The continuous-time scheduling formulation, as introduced in Section 2.5, will produce sub-optimal results if the number of products exceeds the optimal number of products to produce in a prediction horizon. The number of slots is constrained to be equal to the number of products, causing the optimization to always create n production slots and n transitions even in cases in which $<n$ slots would be most economical in the considered horizon for scheduling and control. To eliminate this sub-optimality, an iterative method is introduced to leverage the computational lightness of the MILP continuous-time scheduling formulation. The number of slots in the continuous-time schedule is selected iteratively based on improvement to the objective function (profit), beginning from one slot. As previously mentioned, transition times and control profiles between steady-state products are stored in memory, requiring no computation in online operation. Additionally, the transitions from current measured state to each steady-state product ($\tau_{0'i}$) are calculated once before iterations are initiated. Thus, the iterative method only iterates the MILP problem, not requiring any recalculation of grade transition NLP dynamic optimization problems. This decomposition is computationally light and allows for a fixed-horizon non-cyclic scheduling and control formulation. This non-cyclic fixed-horizon approach to combined scheduling and control enables response to market fluctuations in maximum demand and product price, whereas traditional continuous-time scheduling requires a makespan (T_M) to meet a demand rather than producing an optimal amount of each product within a given fixed horizon. Additional details for this formulation are included in another paper [60].

2.5. Mathematical Formulation

This continuous-time optimization used in phases 1–3 seeks to maximize profit and minimize grade transitions (and associated waste material production) while observing scheduling constraints. The objective function is formulated as follows:

$$\max_{z_{i,s}, t_i^s, t_i^f \forall i,s} \quad J = \sum_{i=1}^{n} \Pi_i \omega_i - \sum_{i=1}^{n} c_{storage,i} \omega_i \sum_{s=1}^{m} z_{i,s}(T_M - t_s^f) - W_\tau \sum_{s=1}^{m} \tau_s, \tag{1}$$

$$\text{s.t.} \quad \text{Equations (2)–(8),}$$

where T_M is the makespan, n is the number of products, m is the number of slots ($m = n$ in these cyclic schedules), $z_{i,s}$ is the binary variable that governs the assignment of product i to a particular slot s, t_s^s is the start time of the slot s, t_s^f is the end time of slot s, Π_i is the per unit price of product i, W_τ is an optional weight on grade transition minimization, τ_s is the transition time within slot s, $c_{storage,i}$ is the per unit cost of storage for product i, and ω_i represents the amount of product i manufactured,

$$\omega_i = \sum_{s=1}^{m} \int_{t_s^s + \tau_s}^{t_s^f} z_{i,s} q \, dt, \tag{2}$$

and where q is the production volumetric flow rate and τ_s is the transition time between the product made in slot $s - 1$ and product i made in slot s. The time points must satisfy the precedence relations:

$$t_s^f > t_s^s + \tau_s \qquad \forall s > 1, \tag{3}$$

$$t_s^s = t_{s-1}^f \qquad \forall s \neq 1, \tag{4}$$

$$t_m^f = T_M, \tag{5}$$

which require that a time slot be longer than the corresponding transition time, impose the coincidence of the end time of one time slot with the start time of the subsequent time slot, and define the relationship between the end time of the last time slot (t_n^f) and the total makespan or horizon duration (T_M).

Products are assigned to each slot using a set of binary variables, $z_{i,s} \in \{0,1\}$, along with constraints of the form:

$$\sum_{s=1}^{m} z_{i,s} = 1 \qquad \forall i, \tag{6}$$

$$\sum_{i=1}^{n} z_{i,s} = 1 \qquad \forall s, \tag{7}$$

which ensure that one product is made in each time slot and each product is produced once.

The makespan is fixed to an arbitrary horizon for scheduling. Demand constraints restrict production from exceeding the maximum demand (δ_i) for a given product, as follows:

$$\omega_i \leq \delta_i \qquad \forall i. \tag{8}$$

The continuous-time scheduling optimization requires transition times between steady-state products ($\tau_{i'i}$) as well as transition times from the current state to each steady-state product if initial state is not at steady-state product conditions ($\tau_{0'i}$).

Transition times are estimated using NMPC via the following objective function:

$$\min_{u} \quad J = (x - x_{sp})^T W_{sp} (x - x_{sp}) + \Delta u^T W_{\Delta u} + u^T W_u, $$
$$\text{s.t.} \quad \text{nonlinear process model} \tag{9}$$
$$x(t_0) = x_0,$$

where W_{sp} is the weight on the set point for meeting target product steady-state, $W_{\Delta u}$ is the weight on restricting manipulated variable movement, W_u is the cost for the manipulated variables, u is the vector of manipulated variables, x_{sp} is the target product steady-state, and x_0 is the start process state from which the transition time is being estimated. The transition time is taken as the time at which and after which $|x - x_{sp}| < \delta$, where δ is a tolerance for meeting product steady-state operating conditions. This formulation harnesses knowledge of nonlinear process dynamics in the system model to find an optimal trajectory and minimum time required to transition from an initial concentration to a desired concentration. This method for estimating transition times also effectively captures the actual behavior of the controller selected, as the transition times are estimated by a simulation of actual controller implementation. This work uses $W_{\Delta u} = 0$ and $W_u = 0$.

3. Case Study Application

As shown in prior work, there are many different strategies for integrating scheduling and control. A novel contribution of this work is a systematic comparison of four general levels of integration through a single case study. In this section, the model and scenarios used to demonstrate progressive economic benefit from the integration of scheduling and control for continuous chemical processes are presented.

3.1. Process Model

This section presents a standard CSTR problem used to highlight the value of the formulation introduced in this work. The CSTR model is applicable in various industries from food/beverage to oil and gas and chemicals. Notable assumptions of a CSTR include:

- Constant volume;
- Well mixed;
- Constant density.

The model shown in Equations (10) and (11) is an example of an exothermic, first-order reaction of $A \Rightarrow B$, where the reaction rate is defined by an Arrhenius expression and the reactor temperature is controlled by a cooling jacket:

$$\frac{dC_A}{dt} = \frac{q}{V}(C_{A0} - C_A) - k_0 e^{-E_A/RT} C_A, \tag{10}$$

$$\frac{dT}{dt} = \frac{q}{V}(T_f - T) - \frac{1}{\rho C_p} k_0 e^{\frac{-E_A}{RT}} C_A \Delta H_r - \frac{UA}{V\rho C_p}(T - T_c). \tag{11}$$

In these equations, C_A is the concentration of reactant A, C_{A0} is the feed concentration, q is the inlet and outlet volumetric flowrate, V is the tank volume (q/V signifies the residence time), E_A is the reaction activation energy, R is the universal gas constant, UA is an overall heat transfer coefficient times the tank surface area, ρ is the fluid density, C_p is the fluid heat capacity, k_0 is the rate constant, T_f is the temperature of the feed stream, C_{A0} is the inlet concentration of reactant A, ΔH_r is the heat of reaction, T is the temperature of reactor and T_c is the temperature of cooling jacket. Table 3 lists the CSTR parameters used.

Table 3. Reactor parameter values.

Parameter	Value
V	100 m^3
E_A/R	8750 K
$\frac{UA}{V\rho C_p}$	2.09 s^{-1}
k_0	7.2×10^{10} s^{-1}
T_f	350 K
C_{A0}	1 mol/L
$\frac{\Delta H_r}{\rho C_p}$	-209 K m^3 mol^{-1}
q	100 m^3/h

In this example, one reactor can make multiple products by varying the concentrations of A and B in the outlet stream. The manipulated variable in this optimization is T_c, which is bounded by 200 K $\leq T_c \leq$ 500 K and by a constraint on manipulated variable movement as $\Delta T_c \leq 2$ K/min.

3.2. Scenarios

The sample problem uses three products over a 24-h horizon. The product descriptions are shown in Table 4, where the product specification tolerance (δ) is ±0.05 mol/L.

Table 4. Product specifications.

Product	C_A (mol/L)	Max Demand (m^3)	Price ($/m^3)	Storage Cost ($/h/m^3)
1	0.10	1000	22	0.11
2	0.30	1000	29	0.1
3	0.50	1000	23	0.12

The transition times between products, as calculated by NMPC using Equation (9), is shown in Table 5.

Table 5. Transition Times Between Products (h).

Starting	Final Product		
Product	1	2	3
1	0	0.50	0.833
2	0.50	0	0.50
3	0.417	0.833	0

Three scenarios are applied to each phase of progressive integration of scheduling and control:

(A) Process disturbance (C_A);
(B) Demand disturbance;
(C) Price disturbance.

Scenarios A–C maintain the specifications in Table 4 but introduce process disturbances, demand disturbances, and price disturbances, respectively (see Table 6). Scenario A introduces a process disturbance to the concentration in the reactor (C_A) of 0.15 mol/L, ramping uncontrollably over 1.4 h. Scenario B introduces a market update with a 20% increase in demand for product 2. Scenario C shows a market update with fluctuations in selling prices for products 2 and 3. The starting concentration for each scenario is 0.10 mol/L, the steady-state product conditions for product 1.

Table 6. Scenario descriptions.

Scenario	Time (h)	Disturbance		
		Product 1	Product 2	Product 3
A	2.2–3.8	———— 0.15 mol/L ————		
B	3.1	+0 m^3	+200 m^3	+0 m^3
C	2.1	+0 $/m^3	−9 $/m^3	+6 $/m^3

4. Results

The results of implementation of each phase for each scenario are discussed and presented in this section. Each problem is formulated in the Pyomo framework for modeling and optimization [61,62]. Nonlinear programming dynamic optimization problems are solved via orthogonal collocation on finite elements [63] with 5 min time discretization and the APOPT and COUENNE MINLP solvers are utilized to solve all mathematical programming problems presented in this work [64,65]. For comparative purposes, profits are compared to those of Phase 3 due to its centrality in performance.

4.1. Scenario A: Process Disturbance

In Scenario A, phase 1 has a poor schedule due to a lack of incorporation of process dynamics into scheduling. The durations of grade transitions, as dictated by process dynamics, are unaccounted for. However, the production amounts or production durations for each product are optimized based on selling prices. The order is selected based on storage costs, clearly leading to longer grade transitions than necessary. The schedule maximizes production of higher-selling products 2 and 3. Phase 1 does not recalculate the schedule after the process disturbance, holding to pre-determined transition timing.

Phase 2 follows the same pattern as phase 1 due to its lack of incorporation of process dynamics. Phase 2 recalculates a schedule after the process disturbance, but because it does not account for process dynamics, it cannot determine that it would be faster to transition to Product 3 from the disturbed process state than to return to Product 2. Thus, the production sequence remains sub-optimal. However, the recalculated schedule enables more profitable Product 2 to be produced than in Phase 1 as the timing of transition to Product 1 is delayed due to the disturbance by the recalculation. Phase 2 illustrates benefit that comes from frequent schedule recalculation rather than from scheduling and control integration.

Phase 3 does not react optimally to the process disturbance because it has a fixed schedule, but its initial schedule is optimal due to the incorporation of process dynamics and the resultant minimization of grade transition durations. Phase 3 illustrates benefit originating solely from scheduling and control integration, without schedule recalculation. Phase 4 optimally reschedules with understanding of transition behavior from the disturbed state to each steady-state operating condition, transitioning to Product 2 immediately after the disturbance. Phase 4 demonstrates the premium benefits of both reactive or frequent rescheduling and from scheduling and control integration.

The simulation results of scenario A are shown in Figure 5 and Table 7.

Table 7. Results: Scenario A.

Phase	Description	Profit		Production (m^3)		
		($)	(%)	Product 1	Product 2	Product 3
1	Segregated, Fixed Schedule	3114	(−38%)	367	858	908
2	Segregated, Reactive Schedule	3942	(−21%)	317	900	992
3	Integrated, Fixed Schedule	4983	(+0%)	308	1000	983
4	Integrated, Reactive Schedule	7103	(+43%)	308	1000	983

Table 8. Results: Scenario B.

Phase	Description	Profit		Production (m^3)		
		($)	(%)	Product 1	Product 2	Product 3
1	Segregated, Fixed Schedule	6033	(−19%)	367	967	908
2	Segregated, Reactive Schedule	7446	(+0.1%)	133	1200	908
3	Integrated, Fixed Schedule	7441	(+0%)	317	1000	992
4	Integrated, Reactive Schedule	8676	(+17%)	308	1200	800

4.2. Scenario B: Market Update Containing Demand Fluctuation

As in Scenario A, the production order for Phases 1 and 2 is sub-optimal due to a lack of incorporation of process dynamics in scheduling, or a lack of integration of scheduling and control. Phase 2 improves performance over Phase 1 by reacting to the market update and producing more profitable Product 2, which had a surge in demand, illustrating again the benefits of reactive scheduling. Phase 3 integrates control with scheduling, resulting in an optimal initial schedule minimizing transition durations. The benefits from integrating scheduling and control (Phase 3) and the benefits of reactive scheduling (Phase 2) are approximately the same in Scenario B, differing in profit by only a negligible amount. However, incorporating both reactive scheduling and scheduling and control integration (Phase 4) leads to a large increase in profits. The initial and recalculated schedules in Phase 4 have optimal production sequence, utilizing process dynamics information to minimize grade transition durations. Additionally, recalculation of the integrated scheduling and control problem after the market update allows for increased production of the highest-selling product, leading to increased profit.

The simulation results of scenario B are shown in Figure 6 and Table 8.

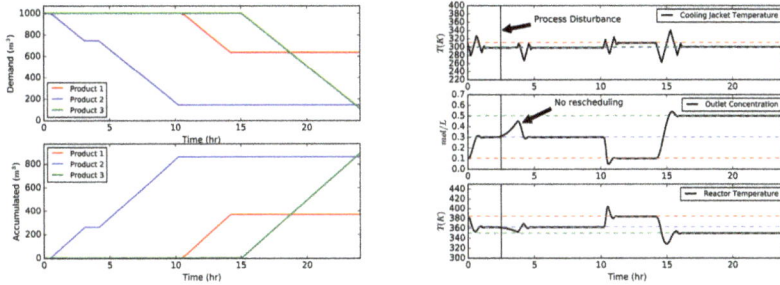

(a) Phase 1: Segregated, Fixed Schedule

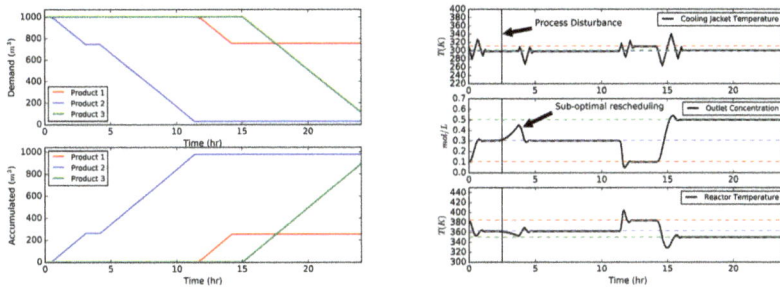

(b) Phase 2: Segregated, Reactive Schedule

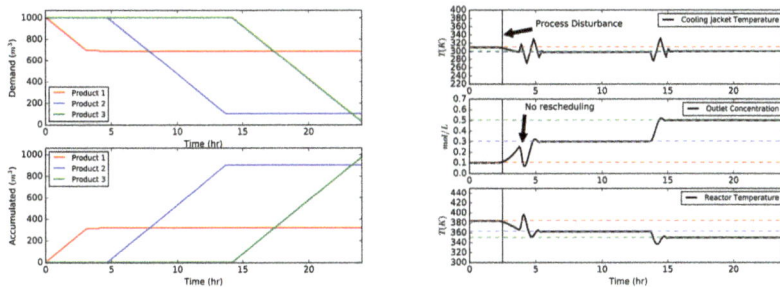

(c) Phase 3: Integrated, Fixed Schedule

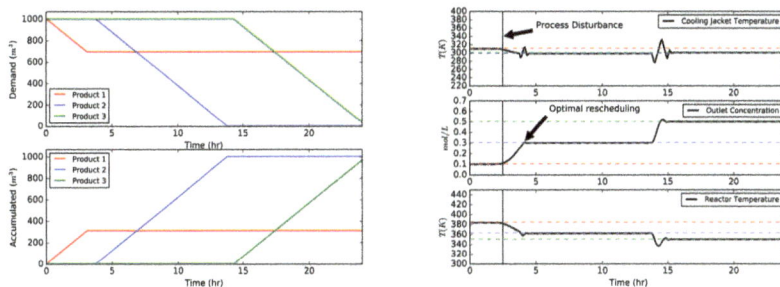

(d) Phase 4: Integrated, Reactive Schedule

Figure 5. Scenario A: Process disturbance.

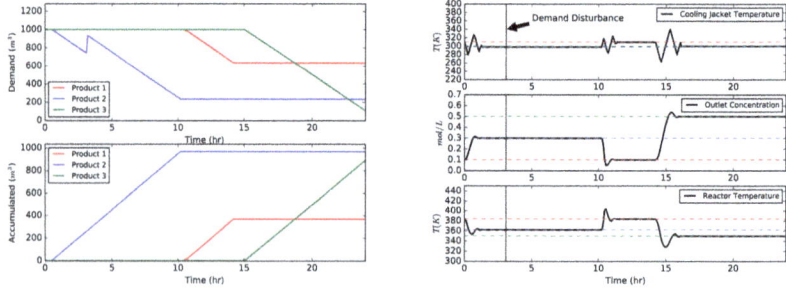

(**a**) Phase 1: Segregated, Fixed Schedule

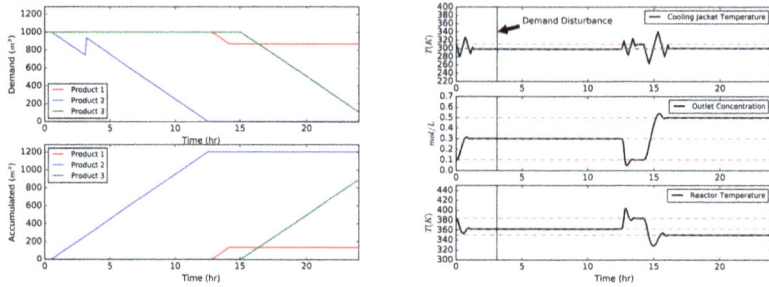

(**b**) Phase 2: Segregated, Reactive Schedule

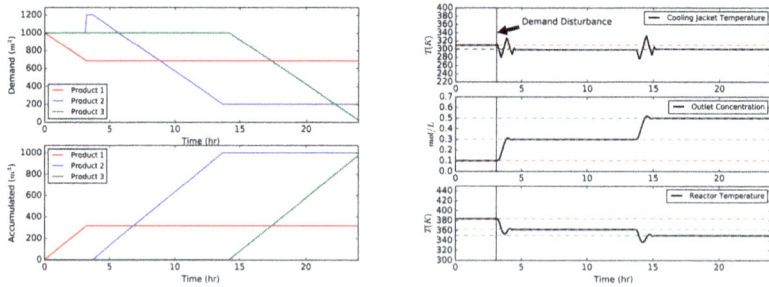

(**c**) Phase 3: Integrated, Fixed Schedule

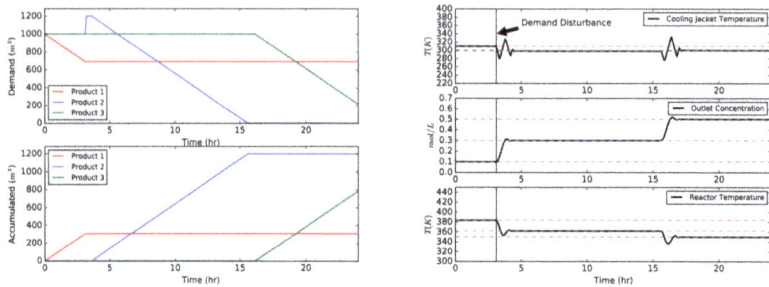

(**d**) Phase 4: Integrated, Reactive Schedule

Figure 6. Scenario B: Market update (demand disturbance).

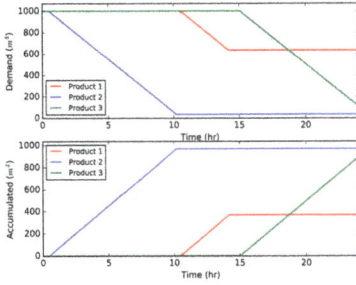

(a) Phase 1: Segregated, Fixed Schedule

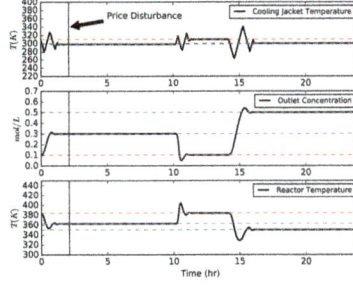

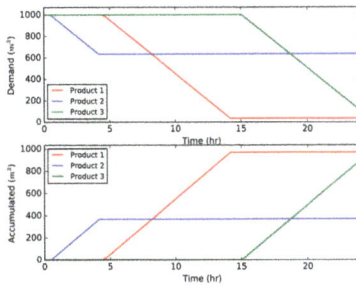

(b) Phase 2: Segregated, Reactive Schedule

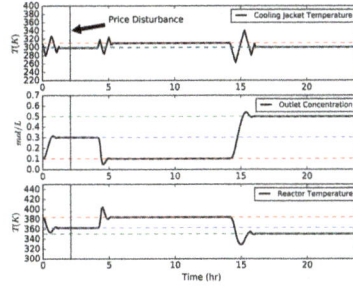

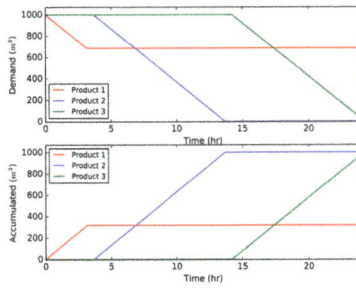

(c) Phase 3: Integrated, Fixed Schedule

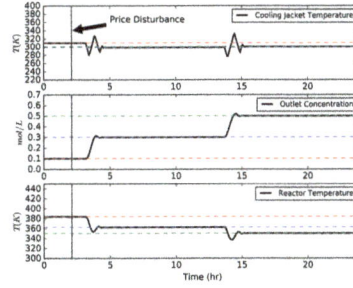

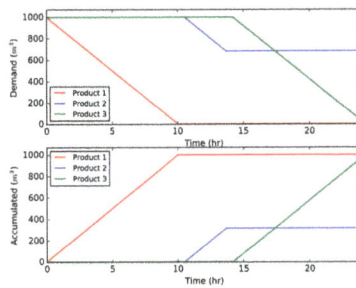

(d) Phase 4: Integrated, Reactive Schedule

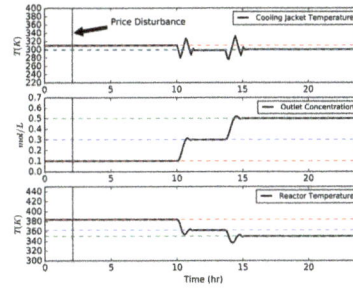

Figure 7. Scenario C: Market update (price disturbance).

4.3. Scenario C: Market Update Containing New Product Selling Prices

As in Scenarios A and B, the production order for Phases 1 and 2 is sub-optimal due to a lack of incorporation of process dynamics in scheduling. However, reactive rescheduling after the price fluctuation information is made available results in a large profit increase from Phase 1 to Phase 2, demonstrating the strength of reactive scheduling even without scheduling and control integration.

Phases 3 and 4 have an optimal production sequence due to the integration of scheduling and control, leading to higher profits than the corresponding segregated phases. This illustrates again the benefits of scheduling and control integration. Like Phase 2, Phase 4 reschedules when the updated market conditions are made available, producing less of product 2 and more of products 1 and 3 due to the price fluctuations. This leads to a leap in profit compared to Phase 3. Phase 4 with both scheduling and control integration and reactive or more frequent scheduling is again the most profitable phase.

The simulation results of scenario C are shown in Figure 7 and Table 9.

Table 9. Results: Scenario C.

Phase	Description	Profit		Production (m³)		
		($)	(%)	Product 1	Product 2	Product 3
1	Segregated, Fixed Schedule	3758	(−16%)	367	967	908
2	Segregated, Reactive Schedule	4879	(+9%)	967	367	908
3	Integrated, Fixed Schedule	4466	(+0%)	317	1000	992
4	Integrated, Reactive Schedule	5662	(+27%)	1000	317	992

5. Conclusions

This work summarizes and reviews the evidence for the economic benefit from scheduling and control integration, reactive scheduling with process disturbances and market updates, and from a combination of reactive and integrated scheduling and control. This work demonstrates the value of combining scheduling and control and responding to process disturbances or market updates by directly comparing four phases of progressive integration through a benchmark CSTR application and three scenarios with process disturbance and market fluctuations. Both ISC and reactice rescheduling show benefit, though their relative benefits are dependent on the situation. More complete integration (applying ISC in closed-loop control, rather than just the scheduling) demonstrates the most benefit.

Directions for Future Work

This work demonstrates the benefit of ISC through four phases of progressive integration using continuous-time scheduling and NMPC on a CSTR case study with three scenarios. This work introduces a benchmark problem with an application (CSTR) and three scenarios on which to benchmark the performance of a scheduling and control formulation. The development of additional benchmark problems applicable to a wider variety of industrial scenarios is proposed as an important potential subject of future work. With increasing research in ISC, benchmark problems for formulation performance comparison of integrated scheduling and control formulations as well as for comparison against a baseline segregated scheduling and control formulation are increasingly important. Benchmark applications and scenarios applicable to batch processes, multi-product continuous processes, and other processes with scenarios representative of probable industrial occurrences should be developed.

This work is applicable to continuous processes considering a single process unit. Progressive integrations proving economic benefit of scheduling and control integration should also be applied to batch processes and continuous processes considering multiple process units. Additionally, this work utilized continuous-time scheduling and NMPC in a decomposed ISC formulation. This formulation inherently considered only steady-state production with no external or dynamic factors (such as time-of-day pricing or dynamic constraints) during production periods. Discrete-time

ISC formulations [18–20,66,67] have been shown to effectively incorporate external and dynamic factors, such as cooling constraints and time-of-day energy pricing. This incorporation enables demand response to time-of-day pricing by reducing or increasing production during periods of steady-state product manufacturing and moving the time of transitions to take advantage of times with relaxed constraints (such as relaxed cooling constraints on exothermic processes). A study of economic benefit of discrete-time formulations as compared to continuous-time formulations for ISC is a potential subject of future work.

Acknowledgments: Financial support from the National Science Foundation Award 1547110 is gratefully acknowledged.

Author Contributions: This paper represents collaborative work by the authors. L.D.R.B., D.P. and G.P. devised the research concepts and strategy. L.D.R.B., D.P., and G.P. implemented formulations and generated simulation data. L.D.R.B. and D.P. analyzed the results. All authors were involved in the preparation of the manuscript.

Conflicts of Interest: The authors declare no conflict of interest.

Abbreviations

The following abbreviations are used in this manuscript:

ISC	integrated scheduling and control
SSC	segregated scheduling and control
MINLP	mixed-integer nonlinear programming
NLP	nonlinear programming
CSTR	continuous stirred tank reactor
MIDO	mixed-integer dynamic optimization
MILP	mixed-integer linear programming
NMPC	nonlinear model predictive control
ASU	air separation unit
MMA	methyl methacrylate reactor
FBR	fluidized bed reactor
RTN	resource task network
HIPS	high impact polystyrene reactor
ASU	cryogenic air separation unit
SISO	single-input single-output
PFR	plug flow reactor
PWA	piecewise affine
DR	demand response

References

1. Backx, T.; Bosgra, O.; Marquardt, W. Integration of Model Predictive Control and Optimization of Processes. In Proceedings of the ADCHEM 2000 International Symposium on Advanced Control of Chemical Processes, Pisa, Italy, 14–16 June 2000; pp. 249–260.
2. Soderstrom, T.A.; Zhan, Y.; Hedengren, J. Advanced Process Control in ExxonMobil Chemical Company: Successes and Challenges. In Proceedings of the AIChE Spring Meeting, Salt Lake City, UT, USA, 7–12 Novembrer 2010; pp. 1–12.
3. Baldea, M.; Harjunkoski, I. Integrated production scheduling and process control: A systematic review. *Comput. Chem. Eng.* **2014**, *71*, 377–390.
4. Nyström, R.H.; Harjunkoski, I.; Kroll, A. Production optimization for continuously operated processes with optimal operation and scheduling of multiple units. *Comput. Chem. Eng.* **2006**, *30*, 392–406.
5. Chatzidoukas, C.; Perkins, J.D.; Pistikopoulos, E.N.; Kiparissides, C. Optimal grade transition and selection of closed-loop controllers in a gas-phase olefin polymerization fluidized bed reactor. *Chem. Eng. Sci.* **2003**, *58*, 3643–3658.
6. Capón-García, E.; Guillén-Gosálbez, G.; Espuña, A. Integrating process dynamics within batch process scheduling via mixed-integer dynamic optimization. *Chem. Eng. Sci.* **2013**, *102*, 139–150.

7. Nie, Y.; Biegler, L.T.; Wassick, J.M. Integrated scheduling and dynamic optimization of batch processes using state equipment networks. *AIChE J.* **2012**, *58*, 3416–3432.
8. Flores-Tlacuahuac, A.; Grossmann, I.E. Simultaneous scheduling and control of multiproduct continuous parallel lines. *Ind. Eng. Chem. Res.* **2010**, *49*, 7909–7921.
9. Terrazas-Moreno, S.; Flores-Tlacuahuac, A.; Grossmann, I.E. Lagrangean heuristic for the scheduling and control of polymerization reactors. *AIChE J.* **2008**, *54*, 163–182.
10. Pattison, R.C.; Touretzky, C.R.; Harjunkoski, I.; Baldea, M. Moving Horizon Closed-Loop Production Scheduling Using Dynamic Process Models. *AIChE J.* **2017**, *63*, 639–651.
11. Engell, S.; Harjunkoski, I. Optimal operation: Scheduling, advanced control and their integration. *Comput. Chem. Eng.* **2012**, *47*, 121–133.
12. Harjunkoski, I.; Maravelias, C.T.; Bongers, P.; Castro, P.M.; Engell, S.; Grossmann, I.E.; Hooker, J.; Méndez, C.; Sand, G.; Wassick, J. Scope for industrial applications of production scheduling models and solution methods. *Comput. Chem. Eng.* **2014**, *62*, 161–193.
13. Harjunkoski, I.; Nyström, R.; Horch, A. Integration of scheduling and control—Theory or practice? *Comput. Chem. Eng.* **2009**, *33*, 1909–1918.
14. Shobrys, D.E.; White, D.C. Planning, scheduling and control systems: Why cannot they work together. *Comput. Chem. Eng.* **2002**, *26*, 149–160.
15. Baldea, M.; Du, J.; Park, J.; Harjunkoski, I. Integrated production scheduling and model predictive control of continuous processes. *AIChE J.* **2015**, *61*, 4179–4190.
16. Baldea, M.; Touretzky, C.R.; Park, J.; Pattison, R.C. Handling Input Dynamics in Integrated Scheduling and Control. In Proceedings of the 2016 IEEE International Conference on Automation, Quality and Testing, Robotics (AQTR), Cluj-Napoca, Romania, 19–21 May 2016; pp. 1–6.
17. Baldea, M. Employing Chemical Processes as Grid-Level Energy Storage Devices. *Adv. Energy Syst. Eng.* **2017**, 247–271, doi:10.1007/978-3-319-42803-1_9.
18. Beal, L.D.R.; Clark, J.D.; Anderson, M.K.; Warnick, S.; Hedengren, J.D. Combined Scheduling and Control with Diurnal Constraints and Costs Using a Discrete Time Formulation. In Proceedings of the FOCAPO/CPC, Tucson, Arizona, 8–12 January 2017.
19. Beal, L.D.; Petersen, D.; Grimsman, D.; Warnick, S.; Hedengren, J.D. Integrated Scheduling and Control in Discrete-Time with Dynamic Parameters and Constraints. *Comput. Chem. Eng.* **2008**, *32*, 463–476.
20. Beal, L.D.; Park, J.; Petersen, D.; Warnick, S.; Hedengren, J.D. Combined model predictive control and scheduling with dominant time constant compensation. *Comput. Chem. Eng.* **2017**, *104*, 271–282.
21. Cai, Y.; Kutanoglu, E.; Hasenbein, J.; Qin, J. Single-machine scheduling with advanced process control constraints. *J. Sched.* **2012**, *15*, 165–179.
22. Chatzidoukas, C.; Kiparissides, C.; Perkins, J.D.; Pistikopoulos, E.N. Optimal grade transition campaign scheduling in a gas-phase polyolefin FBR using mixed integer dynamic optimization. *Comput. Aided Chem. Eng.* **2003**, *15*, 744–747.
23. Chatzidoukas, C.; Pistikopoulos, S.; Kiparissides, C. A Hierarchical Optimization Approach to Optimal Production Scheduling in an Industrial Continuous Olefin Polymerization Reactor. *Macromol. React. Eng.* **2009**, *3*, 36–46.
24. Chu, Y.; You, F. Integration of scheduling and control with online closed-loop implementation: Fast computational strategy and large-scale global optimization algorithm. *Comput. Chem. Eng.* **2012**, *47*, 248–268.
25. Chu, Y.; You, F. Integration of production scheduling and dynamic optimization for multi-product CSTRs: Generalized Benders decomposition coupled with global mixed-integer fractional programming. *Comput. Chem. Eng.* **2013**, *58*, 315–333.
26. Chu, Y.; You, F. Integrated Scheduling and Dynamic Optimization of Sequential Batch Proesses with Online Implementation. *AIChE J.* **2013**, *59*, 2379–2406.
27. Chu, Y.; You, F. Integrated Scheduling and Dynamic Optimization of Complex Batch Processes with General Network Structure Using a Generalized Benders Decomposition Approach. *Ind. Eng. Chem. Res.* **2013**, *52*, 7867–7885.
28. Chu, Y.; You, F. Integration of scheduling and dynamic optimization of batch processes under uncertainty: Two-stage stochastic programming approach and enhanced generalized benders decomposition algorithm. *Ind. Eng. Chem. Res.* **2013**, *52*, 16851–16869.

29. Chu, Y.; You, F. Moving Horizon Approach of Integrating Scheduling and Control for Sequential Batch Processes. *AIChE J.* **2014**, *60*, 1654–1671.

30. Chu, Y.; You, F. Integrated Planning, Scheduling, and Dynamic Optimization for Batch Processes: MINLP Model Formulation and Efficient Solution Methods via Surrogate Modeling. *Ind. Eng. Chem. Res.* **2014**, *53*, 13391–13411.

31. Chu, Y.; You, F. Integrated scheduling and dynamic optimization by stackelberg game: Bilevel model formulation and efficient solution algorithm. *Ind. Eng. Chem. Res.* **2014**, *53*, 5564–5581.

32. Dias, L.S.; Ierapetritou, M.G. Integration of scheduling and control under uncertainties: Review and challenges. *Chem. Eng. Res. Des.* **2016**, *116*, 98–113.

33. Du, J.; Park, J.; Harjunkoski, I.; Baldea, M. A time scale-bridging approach for integrating production scheduling and process control. *Comput. Chem. Eng.* **2015**, *79*, 59–69.

34. Flores-Tlacuahuac, A.; Grossmann, I.E. Simultaneous Cyclic Scheduling and Control of a Multiproduct CSTR. *Ind. Eng. Chem. Res.* **2006**, *45*, 6698–6712.

35. Gutierrez-Limon, M.A.; Flores-Tlacuahuac, A.; Grossmann, I.E. A Multiobjective Optimization Approach for the Simultaneous Single Line Scheduling and Control of CSTRs. *Ind. Eng. Chem. Res.* **2011**, *51*, 5881–5890.

36. Gutierrez-Limon, M.A.; Flores-Tlacuahuac, A.; Grossmann, I.E. A reactive optimization strategy for the simultaneous planning, scheduling and control of short-period continuous reactors. *Comput. Chem. Eng.* **2016**, *84*, 507–515.

37. Gutiérrez-Limón, M.A.; Flores-Tlacuahuac, A.; Grossmann, I.E. MINLP formulation for simultaneous planning, scheduling, and control of short-period single-unit processing systems. *Ind. Eng. Chem. Res.* **2014**, *53*, 14679–14694.

38. Koller, R.W.; Ricardez-Sandoval, L.A. A Dynamic Optimization Framework for Integration of Design, Control and Scheduling of Multi-product Chemical Processes under Disturbance and Uncertainty. *Comput. Chem. Eng.* **2017**, *106*, 147–159.

39. Nie, Y.; Biegler, L.T.; Villa, C.M.; Wassick, J.M. Discrete Time Formulation for the Integration of Scheduling and Dynamic Optimization. *Ind. Eng. Chem. Res.* **2015**, *54*, 4303–4315.

40. Nyström, R.H.; Franke, R.; Harjunkoski, I.; Kroll, A. Production campaign planning including grade transition sequencing and dynamic optimization. *Comput. Chem. Eng.* **2005**, *29*, 2163–2179.

41. Patil, B.P.; Maia, E.; Ricardez-Sandoval, L.A. Integration of Scheduling, Design, and Control of Multiproduct Chemical Processes Under Uncertainty. *AIChE J.* **2015**, *61*, 2456–2470.

42. Pattison, R.C.; Touretzky, C.R.; Johansson, T.; Harjunkoski, I.; Baldea, M. Optimal Process Operations in Fast-Changing Electricity Markets: Framework for Scheduling with Low-Order Dynamic Models and an Air Separation Application. *Ind. Eng. Chem. Res.* **2016**, *55*, 4562–4584.

43. Prata, A.; Oldenburg, J.; Kroll, A.; Marquardt, W. Integrated scheduling and dynamic optimization of grade transitions for a continuous polymerization reactor. *Comput. Chem. Eng.* **2008**, *32*, 463–476.

44. Terrazas-Moreno, S.; Flores-Tlacuahuac, A.; Grossmann, I.E. Simultaneous design, scheduling, and optimal control of a methyl-methacrylate continuous polymerization reactor. *AIChE J.* **2008**, *54*, 3160–3170.

45. Terrazas-Moreno, S.; Flores-Tlacuahuac, A.; Grossmann, I.E. Simultaneous cyclic scheduling and optimal control of polymerization reactors. *AIChE J.* **2007**, *53*, 2301–2315.

46. You, F.; Grossmann, I.E. Design of responsive supply chains under demand uncertainty. *Comput. Chem. Eng.* **2008**, *32*, 3090–3111.

47. Zhuge, J.; Ierapetritou, M.G. Integration of Scheduling and Control with Closed Loop Implementation. *Ind. Eng. Chem. Res.* **2012**, *51*, 8550–8565.

48. Zhuge, J.; Ierapetritou, M.G. Integration of Scheduling and Control for Batch Processes Using Multi-Parametric Model Predictive Control. *AIChE J.* **2014**, *60*, 3169–3183.

49. Zhuge, J.; Ierapetritou, M.G. An Integrated Framework for Scheduling and Control Using Fast Model Predictive Control. *AIChE J.* **2015**, *61*, 3304–3319.

50. Zhuge, J.; Ierapetritou, M.G. A Decomposition Approach for the Solution of Scheduling Including Process Dynamics of Continuous Processes. *Ind. Eng. Chem. Res.* **2016**, *55*, 1266–1280.

51. Mahadevan, R.; Doyle, F.J.; Allcock, A.C. Control-relevant scheduling of polymer grade transitions. *AIChE J.* **2002**, *48*, 1754–1764.

52. Mojica, J.L.; Petersen, D.; Hansen, B.; Powell, K.M.; Hedengren, J.D. Optimal combined long-term facility design and short-term operational strategy for CHP capacity investments. *Energy* **2017**, *118*, 97–115.

53. Gupta, D.; Maravelias, C.T.; Wassick, J.M. From rescheduling to online scheduling. *Chem. Eng. Res. Des.* **2016**, *116*, 83–97.

54. Gupta, D.; Maravelias, C.T. On deterministic online scheduling: Major considerations, paradoxes and remedies. *Comput. Chem. Eng.* **2016**, *94*, 312–330.

55. Gupta, D.; Maravelias, C.T. A General State-Space Formulation for Online Scheduling. *Processes* **2017**, *4*, 69.

56. Kopanos, G.M.; Pistikopoulos, E.N. Reactive scheduling by a multiparametric programming rolling horizon framework: A case of a network of combined heat and power units. *Ind. Eng. Chem. Res.* **2014**, *53*, 4366–4386.

57. Liu, S.; Shah, N.; Papageorgiou, L.G. Multiechelon Supply Chain Planning With Sequence-Dependent Changeovers and Price Elasticity of Demand under Uncertainty. *AIChE J.* **2012**, *58*, 3390–3403.

58. Touretzky, C.R.; Baldea, M. Integrating scheduling and control for economic MPC of buildings with energy storage. *J. Process Control* **2014**, *24*, 1292–1300.

59. Li, Z.; Ierapetritou, M.G. Process Scheduling Under Uncertainty Using Multiparametric Programming. *AIChE J.* **2007**, *53*, 3183–3203.

60. Petersen, D.; Beal, L.D.R.; Prestwich, D.; Warnick, S.; Hedengren, J.D. Combined Noncyclic Scheduling and Advanced Control for Continuous Chemical Processes. *Processes* **2017**, *4*, 83, doi:10.3390/pr5040083.

61. Hart, W.E.; Watson, J.P.; Woodruff, D.L. Pyomo: Modeling and solving mathematical programs in Python. *Math. Program. Comput.* **2011**, *3*, 219–260.

62. Hart, W.E.; Laird, C.; Watson, J.P.; Woodruff, D.L. *Pyomo—Optimization Modeling in Python*; Springer Science+Business Media, LLC: Berlin/Heidelberg, Germany, 2012; Volume 67.

63. Carey, G.; Finlayson, B.A. Orthogonal collocation on finite elements for elliptic equations. *Chem. Eng. Sci.* **1975**, *30*, 587–596.

64. Hedengren, J.; Mojica, J.; Cole, W.; Edgar, T. APOPT: MINLP Solver for Differential Algebraic Systems with Benchmark Testing. In Proceedings of the INFORMS Annual Meeting, Pheonix, AZ, USA, 14–17 October 2012.

65. Belotti, P.; Lee, J.; Liberti, L.; Margot, F.; Wächter, A. Branching and bounds tightening techniques for non-convex MINLP. *Optim. Methods Softw.* **2009**, *24*, 597–634.

66. Floudas, C.A.; Lin, X. Continuous-time versus discrete-time approaches for scheduling of chemical processes: A review. *Comput. Chem. Eng.* **2004**, *28*, 2109–2129.

67. Sundaramoorthy, A.; Maravelias, C.T. Computational study of network-based mixed-integer programming approaches for chemical production scheduling. *Ind. Eng. Chem. Res.* **2011**, *50*, 5023–5040.

![processes logo]

MDPI

Article

Combined Noncyclic Scheduling and Advanced Control for Continuous Chemical Processes

Damon Petersen [1], Logan D. R. Beal [1], Derek Prestwich [1], Sean Warnick [2] and John D. Hedengren [1,*]

[1] Department of Chemical Engineering, Brigham Young University, Provo, UT 84602, USA; damon.chem.e@gmail.com (D.P.); beall@byu.edu (L.D.R.B.); derekprestwich@hotmail.com (D.P.)
[2] Department of Computer Science, Brigham Young University, Provo, UT 84602, USA; sean.warnick@gmail.com
* Correspondence: john.hedengren@byu.edu

Received: 14 November 2017; Accepted: 8 December 2017; Published: 13 December 2017

Abstract: A novel formulation for combined scheduling and control of multi-product, continuous chemical processes is introduced in which nonlinear model predictive control (NMPC) and noncyclic continuous-time scheduling are efficiently combined. A decomposition into nonlinear programming (NLP) dynamic optimization problems and mixed-integer linear programming (MILP) problems, without iterative alternation, allows for computationally light solution. An iterative method is introduced to determine the number of production slots for a noncyclic schedule during a prediction horizon. A filter method is introduced to reduce the number of MILP problems required. The formulation's closed-loop performance with both process disturbances and updated market conditions is demonstrated through multiple scenarios on a benchmark continuously stirred tank reactor (CSTR) application with fluctuations in market demand and price for multiple products. Economic performance surpasses cyclic scheduling in all scenarios presented. Computational performance is sufficiently light to enable online operation in a dual-loop feedback structure.

Keywords: scheduling; model predictive control; dynamic market; process disturbances; nonlinear

1. Introduction

Production scheduling and advanced control are terms which describe efforts to optimize chemical manufacturing operations. Production scheduling seeks to optimally pair production resources with production demands to maximize operational profit. Advanced controls seek to optimally control a chemical process to observe environmental and safety constraints and to drive operations to the most economical conditions. Model predictive advanced controls use process models to make predictions into a future horizon based on possible control moves to determine the optimal sequence of control moves to meet an objective, such as reaching an operational set point. In a multi-product continuous chemical process, steady-state operational set points or desired operational conditions over a future time horizon are determined by the production schedule, determining at which times, in what amounts, and in which sequence certain products should be produced.

As scheduling and advanced control are closely interrelated, both seeking to optimize chemical manufacturing efficiency over future time horizons, their integration has been the subject of significant recent investigation. Multiple review articles have been published on the integration of scheduling and control [1–4]. As schedules are informed of process dynamics as dictated by control structure and process nonlinearities, schedules produced become more aligned with actual process operations and schedule efficacy improves [5]. Conversely, when scheduling and advanced control are separated, coordination of closed-loop responses to process disturbances is lost, unrealistic set points may be passed from scheduling to advanced controls, and advanced control may seek to drive the process to sub-optimal operating conditions due to a lack of communication [3,5,6]. In the presence of a process disturbance, for example, advanced controls may attempt to return to a set point determined prior to the disturbance, whereas a recalculation of the schedule from the measured disturbed state, with knowledge of process dynamics and controller behavior, may show a different schedule and set point to be most economical [5,7].

One setback to the integration of scheduling and control is computational difficulty. Advanced controls, particularly model predictive controls, utilize dynamic process models in dynamic optimization problems, forming linear programming (LP) or nonlinear programming (NLP) optimization problems depending on the type of model used. Scheduling involves discrete or binary decisions, such as assigning particular products to production at given times. This gives rise to mixed-integer programming problems for scheduling. When scheduling and control are combined, the computational burden of mixed-integer programming is combined with the LP or NLP dynamic optimization problems. Additionally, dynamic optimization control problems are not required to be solved only once, as in an iteration of advanced online control, but for each grade transition during a production schedule. The integrated scheduling and control (ISC) problem has been shown to be computationally difficult, and much research has been invested in decomposition and reduction of computational burden for the problem [8–12].

Reduction of the computational burden of integrated scheduling and control is especially important for enabling online implementations. It has been shown repeatedly in simulation that closed-loop online implementations of integrated scheduling and control are critical to recalculating optimal scheduling and control when faced with process disturbances [5,7]. Additionally, it has been demonstrated that online closed-loop integrated scheduling and control is vital to optimally responding to variable market conditions, including demand fluctuations and price fluctuations [5,13,14]. As mentioned in review articles on integrated scheduling and control, a key motivator for integration is the reduced time-scale at which market conditions fluctuate [1,2]. A reduced time-scale for market fluctuations implies that the time-scale at which the economically optimal schedule and associated optimal control profile fluctuates is likewise reduced. Thus, online recalculation to respond to market condition updates is critical to integrated scheduling and control, and computational burden reduction to enable such implementation is a salient topic for researchers.

Online responsiveness to volatile market conditions can improve process economics by updating or changing the existing schedule [5]. The majority of integrated scheduling and control formulations have used cyclic schedules [7–11,15–22]. However, it has been suggested that a *dynamic* cyclic schedule may improve process economics [23]. Beal et al. suggest that a dynamic cyclic schedule can increase the flexibility of scheduling beyond the rigidity of a cyclic product grade wheel. A dynamic cyclic schedule can dynamically change the sequence and duration of production of products on the grade wheel based on process disturbances, sudden surges in demand for specific products, or time-dependent constraints such as operator availability for equipment handling. Economic benefit from dynamic cyclic scheduling has been demonstrated in previous work [5,7]. Recent developments in integrated scheduling and control with discrete representations of time do away with the idea of a cyclic schedule as the schedule is determined as a sum total of the binary production variables at each discrete point in time during a prediction horizon [24]. In these works, the number of products manufactured, the selection of manufactured products, sequence, and timing are all solved simultaneously with process dynamic models, resulting in a computationally heavy formulation. Evidence of benefits from noncyclic scheduling has also been demonstrated in other fields, such as cluster-tool robot scheduling for semiconductor fabrication [25–30].

This work explores fully noncyclic scheduling integrated with advanced process control. A continuous representation of time is utilized for the scheduling model, and nonlinear model predictive control (NMPC) is utilized for advanced process control. A decomposition is utilized to break the problem into NLP problems and mixed-integer linear programming (MILP) problems without the need for alternating iterations. Further decomposition into problems calculated offline from information known a priori with results stored in memory and problems solved online further reduce the computational burden of integrated scheduling and control, making the problem feasible for online, closed-loop implementation. Noncyclic scheduling is implemented through an iterative method, which iterates through potential numbers of products to produce during a future horizon. The integrated scheduling and control algorithm selects the number of products to manufacture, which products to manufacture, the production sequence, production timing and amounts, as well as the grade transitions between each product. This allows greater flexibility than cyclic scheduling for responding to variable product demands and prices.

2. Literature Review

Extensive prior research has been conducted on integrated scheduling and control, including several literature review articles [1–4,31,32]. Mahadevan et al. incorporate control level considerations such as grade transition costs in production scheduling [33]. Grossman and Flores-Tlacuahuac investigate simultaneous cyclic scheduling and process control for continuously stirred tank reactors (CSTR), polymer manufacturing processes, and for parallel continuous chemical processes [16,17,34]. A mixed-integer dynamic optimization (MIDO) approach has been used to solve the simultaneous continuous-time scheduling and optimal control problem with orthogonal collocation on finite elements. The large MIDO problem is initialized with dynamic optimization grade transition problem solutions [17]. Chatzidoukas et al. study solving the simultaneous dynamic optimization optimal grade transitions and scheduling problem for fluidized-bed reactors (FBR) via a MIDO approach with orthogonal collocation on finite elements [35]. They also study simultaneous selection of closed-loop controllers and selection of controller parameters with the scheduling problem for polymerization processes [36,37]. Terrazas-Moreno and Flores-Tlacuahuac also investigate simultaneous cyclic scheduling and control of continuous chemical processes, and study a langrangean heuristic method for reduction of computational burden [15,18]. Baldea et al. and Du et al. demonstrate reduction of computational burden of the integrated problem by using reduced order, or scale-bridging models (SBM), in scheduling [8,21]. Scale-bridging models encapsulate the core information about process dynamics in a low-order model for integrated solution with scheduling [8,9,21]. Nie et al. study the short-term scheduling and dynamic optimization of multi-unit chemical processes using a state equipment network (SEN) and mixed-logic dynamic optimization (MLDO), solved by a Big M reformulation and orthogonal collocation into an MINLP problem [38]. Nie et al. also study combined scheduling and control of parallel reactors using a resource task network (RTN) coupled with dynamic models [39]. The solution is achieved by iteration between MILP and NLP subproblems and Benders' decomposition. Gutierrez-Limon et al. develop a multi-objective optimization approach for simultaneous scheduling and control of continuous chemical processes [40]. Benefit is demonstrated from using multi-objective approaches with Pareto fronts rather than combining economic and dynamic performance objectives into a single objective function. Gutierrez-Limon and Flores-Tlacuahuac also investigate simultaneous planning, scheduling, and control for continuous chemical processes using a large MINLP problem and NMPC [41]. Prata et al. study the use of outer approximation methods for the solution of integrated scheduling and control problems [42].

Zhuge and Ierapetritou investigate closed-loop implementation of integrated scheduling and control. The benefit of integrated scheduling and control responsive to process disturbances in continuous chemical processes is demonstrated [7]. Further work by Zhuge and Ierapetritou investigates various methodologies for the reduction of computational burden for integrated scheduling and control to enable feasible online implementation. Application of multi-parametric model predictive control (mp-MPC) to shift the solution of optimal control dynamic optimization problems offline while using only an explicit control solution with minimal computational requirements is studied [22]. The explicit control solution from mp-MPC is integrated with scheduling to reduce the computational requirements of the online problem. A simplified piece-wise affine (PWA) model to represent process dynamics for the scheduling level, rather than a first-principles process model, and fast model predictive control (fast MPC) at the control level are implemented, resulting in a significant reduction of computational burden. Additionally, a decomposition approach is presented based on optimality analysis showing that production sequence and transition times are independent of product demands [20]. The transition stages, a smaller-scale mixed-integer nonlinear programming (MINLP) problem, are solved separately from the production stages, a smaller-scale NLP problem, as opposed to solving an integrated large-scale MINLP problem. No need for iterative alternation is necessary between the two smaller scale problems.

Chu and You also investigate closed-loop implementations of integrated scheduling and control. They investigate solutions through mixed-integer fractional programming (MINFP) with a fast, global optimization algorithm in a simultaneous controller selection and scheduling problem for online implementation [10,11]. They also investigate multi-objective optimization, including decomposition into an offline solution of the Pareto frontier and an online MINLP problem simultaneously selecting a batch scheduling recipe and optimal control [43]. Chu and You also investigate two-stage stochastic programming to confront uncertainty in integrated scheduling and control problems; however, the problem is computationally difficult [44]. Reduction of computational difficulty via nested Benders' decomposition is investigated. They also investigate a decomposition into a master scheduling problem (MILP) and a primal problem containing multiple separable dynamic optimization (or control) problems [45]. The primal and master problems are solved together via iterative alternation until convergence to an acceptable error is achieved. Nystrom et al. also examine the use of iterations between master MILP problems and primal NLP problems until convergence to an acceptable error is achieved for the solution of integrated scheduling and control problems [46,47]. Chu and You also investigate integration of planning, scheduling, and control in a large MIDO problem discretized into a MINLP problem by orthogonal collocation on finite elements. They use surrogate models to effectively reduce the computational burden for the large problem. A decomposition into a leading scheduling problem and following dynamic optimization problems based on Stakelberg game theory is also investigated [48].

Rossi has recently demonstrated a decomposition of integrated scheduling and control for batch processes [49]. The schedule is calculated offline but is updated in real time by updates from an online optimal control problem. Bauer et al. has recently suggested that key performance indicators (KPI) are already implemented in industry as the integrating factor between scheduling and control. Scheduling and control for economic model predictive control (EMPC) of buildings with energy storage has been investigated [50].

Integration of design, scheduling, and control has also been investigated. Terrazas-Moreno and Flores-Tlacuahuac examine combined design, scheduling, and control with dynamic process models nearly a decade ago [51]. Koller et al. present an optimization framework for integrated scheduling, control, and design of multi-product processes with uncertainty. Worst-case scenarios are examined in the algorithm to design a process feasible under all worst-case scenarios [52]. Grade transitions and scheduling are incorporated via flexible finite elements and dynamic process models. Patil et al. study combined design, scheduling, and control for continuous processes [53]. They study worst-case scenarios through frequency response analysis and apply their formulation to a high impact polystyrene (HIPS) polymerization process.

Moving horizon implementations of closed-loop integrated scheduling and control have been studied by various researchers. Chu and You use a moving future horizon for online recalculation of integrated scheduling and control for batch processes in the presence of disturbances such as unit breakdowns and system disruptions and a dual feedback structure to reduce the frequency of integrated problem solution [12]. Pattison et al. investigate using a scale-bridging model (SBM) in a closed-loop moving horizon implementation of integrated scheduling and control for an industrial air separation unit (ASU), demonstrating effective response to planned maintenance, unplanned maintenance, and random failure scenarios [14]. A heuristic reactive strategy for integrated planning, scheduling, and control for multi-product continuous chemical processes has been developed and applied in scenarios with market updates with fluctuating product demands [54].

3. Problem Formulation

This work investigates integration of scheduling and control with noncyclic scheduling and NMPC. The algorithm is allowed to select the number of products to manufacture during a prediction horizon, which products to manufacture, the production sequence, production durations, and optimal control moves. The mathematical formulation is presented in this section.

3.1. Decomposition

An objective of this work is feasibility for closed-loop online implementation. To accomplish this objective, the problem is decomposed into an MILP programming problem and multiple separable NLP dynamic optimization, or optimal control, problems without the need for alternating iterations. This formulation builds on previous work that demonstrates the separability of the integrated scheduling and control problem into subproblems without the need for iterations [20] and builds on previous work which demonstrates the separation into MILP and dynamic optimization problems [10,46,47]. This formulation also builds on previous work that demonstrates benefits from shifting separable computational burden into offline portions of the integrated problem [11,19,22,55].

The main decomposition employed in this work is a decomposition into an MILP problem that determines the production sequence and production amounts for each manufactured product, and a group of separable dynamic optimization NLP problems for optimal control during grade transitions. The formulation for the MILP problem is as follows:

$$\max_{\omega_i, z_{i,s}, t_i^s, t_i^f \forall i,s} \quad J = \sum_{i=1}^{n} \Pi_i \omega_i - \sum_{i=1}^{n} c_{storage,i} \omega_i \sum_{s=1}^{m} z_{i,s}(T_M - t_s^f) - W_\tau \sum_{s=1}^{m} \tau_s, \tag{1}$$

$$\text{s.t.} \quad \text{Equations (2)–(8),}$$

where T_M is the makespan, n is the number of products, m is the number of slots (constrained to equal the number of products such that $m = n$), $z_{i,s}$ is the binary variable that governs the assignment of product i to a particular slot s, t_s^s is the start time of the slot s, t_s^f is the end time of slot s, Π_i is the per unit price of product i, W_τ is an optional weight on grade transition minimization, τ_s is the transition time within slot s, $c_{storage,i}$ is the per unit cost of storage for product i, and ω_i represents the amount of product i manufactured,

$$\omega_i = \sum_{s=1}^{m} \int_{t_s^s + \tau_s}^{t_s^f} z_{i,s} q \, dt. \tag{2}$$

Products are assigned to each slot using a set of binary variables, $z_{i,s} \in \{0,1\}$ that assign a product i to be produced in a given slot s. Constraints of the following form are added:

$$\sum_{i=1}^{n} z_{i,s} = 1 \quad \forall s, \tag{3}$$

which limit the assignment of production within each slot, ensuring that one and only one product is assigned to each slot, and

$$\sum_{s=1}^{m} z_{i,s} \leq 1 \quad \forall i, \tag{4}$$

which constrains the production of products. This constraint, unlike traditional continuous-time scheduling formulations, does not require each product to be produced once and only once during the schedule, but allows the optimization to select the number of products to produce during the horizon for scheduling provided each slot has a production assignment (Equation (3)). Not every available product must be produced during each schedule.

The values of the vector τ_s are determined by the optimization variables $z_{i,s}$. τ_s represents the transition time between the product made in slot $s - 1$ and product made in slot s. Thus, the value of τ_s is determined by the optimal values of $z_{j,s-1}$ and $z_{k,s}$ which determine which products are assigned to slots $s - 1$ and s. For $z_{j,s-1}$ and $z_{k,x}$, τ_s would be equal to the calculated grade transition time from product j to product k. The possible values for τ_s are the grade transition times calculated via NLP optimal grade transition problems (Equation (9)). The time points must satisfy the precedence relations:

$$t_s^f > t_s^s + \tau_s \qquad \forall s > 1, \tag{5}$$

$$t_s^s = t_{s-1}^f \qquad \forall s \neq 1, \tag{6}$$

$$t_m^f = T_M, \tag{7}$$

which require that a time slot be longer than the corresponding transition time, impose the coincidence of the end time of one time slot with the start time of the subsequent time slot, and define the relationship between the end time of the last time slot (t_n^f) and the total makespan or horizon duration (T_M).

The makespan is fixed to an arbitrary horizon for scheduling. Demand constraints restrict production from exceeding the maximum demand (δ_i) for a given product, as follows:

$$\omega_i \leq \delta_i \qquad \forall i. \tag{8}$$

The continuous-time scheduling optimization (or MILP problem) requires transition times between steady-state products ($\tau_{i'i}$) as well as transition times from the current state to each steady-state product if initial state is not at steady-state product conditions ($\tau_{0'i}$). These grade transitions comprise the separable dynamic optimization problems or NLP portion of the overall problem decomposition. Grade transition profiles are optimized using the following objective:

$$\min_u \quad J = (x - x_{sp})^T W_{sp}(x - x_{sp}) + \Delta u^T W_{\Delta u} + u^T W_u,$$

$$\text{s.t.} \quad \text{nonlinear process model}, \tag{9}$$

$$x(t_0) = x_0,$$

where W_{sp} is the weight on the set point for meeting target product steady-state, $W_{\Delta u}$ is the weight on restricting manipulated variable movement, W_u is the cost for the manipulated variables, u is the vector of manipulated variables, x_{sp} is the target product steady-state, and x_0 is the start process state from which the transition time is being estimated. The transition time is taken as the time at which and after which $|x - x_{sp}| < \epsilon$, where ϵ is a tolerance for meeting product steady-state operating conditions. This formulation harnesses knowledge of nonlinear process dynamics in the system model to find an optimal trajectory and minimum time required to transition from an initial concentration to a desired concentration. The usage of NMPC for grade transitions in the integrated problem effectively captures the actual behavior of the controller used in the process, as the transition times are estimated by a simulation of actual controller implementation on a first-principles nonlinear model of the process.

Because product steady-state operating conditions can be known a priori, all grade transition times between production steady-state operating conditions can be calculated offline and stored in memory in a grade transition time table: τ_{ss}. The online portion of the NLP subproblem is comprised only of the calculation of τ_θ, the transition duration and corresponding optimal control profile from current measured state (x_θ) and each steady-state operating condition, or, in other words, the vector of possible transitions for the first slot ($s = 0$). For example, consider the case of three products as described in Table 1. As product steady-state conditions are known a priori as shown in Table 1, the transition times between product steady-states can be calculated through NLP problems (Equation (9)) and stored in a grade transition time table prior to operation (Table 2). However, before operation, process state measurements (x_θ) cannot be known. For example, if in the three product case described in Table 1 a process disturbance was measured at $x_\theta = 0.19$ mol/L during online operation, for an optimal rescheduling beginning from current state x_θ to be calculated, transition durations from x_θ to each operating steady-state would need to be estimated online by solving n separable NLP optimal grade transition problems, where n is the number of steady-state products considered. The results of these online optimal transition problems would be the vector τ_θ (Table 3). These transitions would be the possible values for τ_s for the first slot ($s = 0$) in Equation (1) for the MILP rescheduling problem.

Table 1. Product specifications.

Product	C_A (mol/L)
1	0.10
2	0.30
3	0.50

Table 2. Transition time table (τ_{ss}) *.

Start	End Product		
Product	1	2	3
1	0.00	0.71	1.20
2	0.45	0.00	0.71
3	0.94	1.57	0.00

* Transition times in hours.

Table 3. Initial transitions (τ_θ).

Product	τ_θ (h)
1	0.31
2	0.43
3	0.96

With τ_{ss} and τ_θ grade transition information, the MILP problem is equipped to optimally select the production sequence and amounts for the prediction horizon based on product demands, prices, transition durations, raw material cost, storage cost, and other economic parameters. Even when a process disturbance is encountered and measured, the schedule can be optimally recalculated from the measured disturbance, x_θ, via the incorporation of transitions from x_θ to each production steady-state condition (τ_θ).

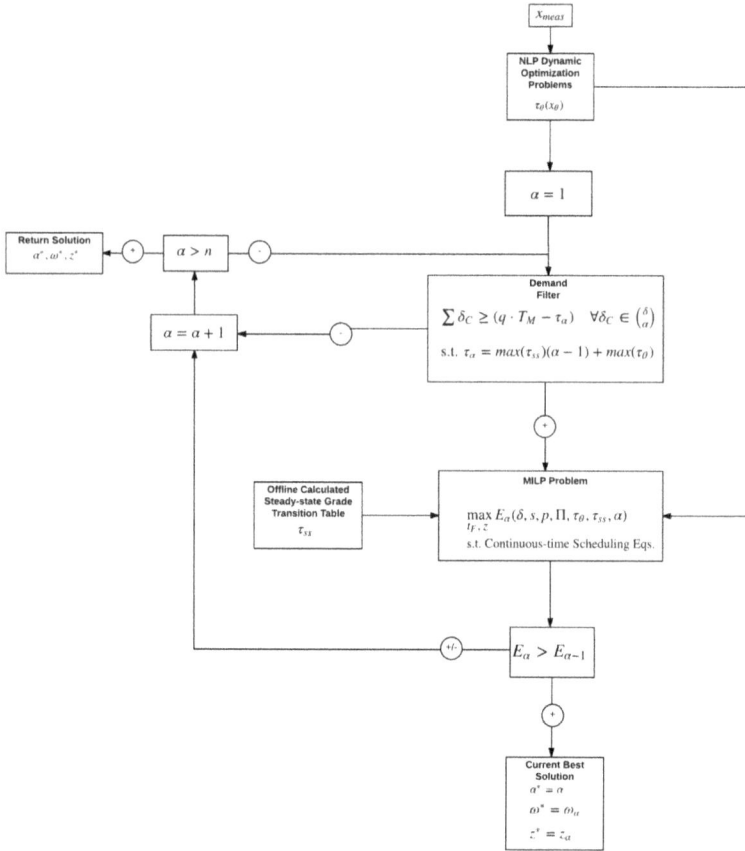

Figure 1. Diagram describing the integrated noncyclic scheduling and optimal control problem. For variable descriptions, see Table 4.

3.2. Iterative Method

The cyclic continuous-time scheduling formulation is sub-optimal if the number of products on the wheel exceeds the optimal number of products to produce in a prediction horizon. The number of slots in the MILP problem is constrained to be equal to the number of products, causing the optimization to always create n production slots and n transitions even in cases in which $<n$ slots would be most economical in the considered horizon for scheduling and control. To allow greater flexibility, an iterative method is introduced which leverages the computational lightness of the separated MILP subproblem. The number of slots in the continuous-time schedule is selected iteratively based on improvement to the objective function (profit), beginning from one slot, or only one product to produce during the horizon (Figure 1). The iterations begin from one slot ($\alpha = 1$) due to the nature of grade transitions. Reducing the number of production slots reduces the number of necessary grade transitions, and reduces the corresponding amount of waste material produced. All grade transition NLP subproblems are calculated prior to slot iterations as grade transition information is independent of the number of production slots. This iterative method enables a noncyclic approach to combined scheduling and control and enables response to market fluctuations in product maximum demand and product price.

A heuristic demand filter is introduced to further reduce computational burden (Figure 1). The filter checks for any combination of α products with summed demand sufficient to fill the prediction horizon for scheduling and control. This effectively filters each possible α for sufficient demand before the MILP problem is executed. In calculating the minimum summed demand which must be met for the given iteration for the MILP problem to be executed, the filter accounts for transition times calculated by NLP dynamic optimization problems both offline (all $\tau_{i,i}$ stored in τ_{ss}) and online (all τ_0 stored in τ_θ). The demand must be sufficient to fill the production potential of the prediction horizon, which equates to $(q \cdot T_M - \tau_\alpha)$, where τ_α is the summed grade transition durations. To ensure each α which has potential for sufficient demand is tested in the MILP problem, the demand filter overestimates for transition times. In the case of underestimation of total transition times (τ_α), the demand required to fill production during the prediction horizon $(q \cdot T_M - \tau_\alpha)$ will be overestimated. In a such case, a combination δ_C which would have had enough demand may inaccurately be deemed insufficient. Since often the optimal number of slots will be the smallest number of slots with sufficient demand (because the number of grade transitions and the corresponding amount of off-specification production increases with the number of slots), it is pertinent to not underestimate the durations of grade transitions. To eliminate such underestimation, the maximum transitions are used for estimations of the total transition time during the prediction horizon (τ_α).

Table 4. Variable descriptions.

Variable	Description
α	Current number of slots
τ_{ss}	Matrix of grade transition durations between production steady-states
x_θ	Measured process state
τ_θ	Vector of grade transition durations from x_θ to each product steady-state
p	Vector of production steady-states known a priori
T_M	Prediction horizon duration or makespan
q	Process flow rate (m^3/h)
τ_α	Estimation of total grade transition during a prediction horizon for α production slots
δ	Vector of maximum demands (δ_i) for products
δ_C	Any combination of α product demands
n	Number of possible products
Π	Vector of product selling prices
s	Vector of product storage costs (m^3/h)
E_α	Estimated profit from optimized schedule for α production slots
ω_α	Vector of manufactured amount per product (m^3) in optimized schedule with α slots

4. Case Study Application

In this section, the performance of the iterative noncyclic integrated scheduling and control formulation in this work is demonstrated. The test scenarios and computational, economic, and closed-loop performance results of the algorithm are presented.

4.1. Process Model

This section presents the CSTR problem used to highlight the value of the formulation introduced in this work. The CSTR model is a general test problem that is widely applicable in various industries from petrochemicals to food processing. The model shown in Equations (10) and (11) is an example of an exothermic, first-order reaction of $A \Rightarrow B$ where the reaction rate is defined by an Arrhenius expression and the reactor temperature is controlled by a cooling jacket:

$$\frac{dC_A}{dt} = \frac{q}{V}(C_{A0} - C_A) - k_0 e^{-E_A/RT} C_A, \tag{10}$$

$$\frac{dT}{dt} = \frac{q}{V}(T_f - T) - \frac{1}{\rho C_p} k_0 e^{\frac{-E_A}{RT}} C_A \Delta H_r - \frac{UA}{V \rho C_p}(T - T_c). \tag{11}$$

In these equations, C_A is the concentration of reactant A, C_{A0} is the feed concentration, q is the inlet and outlet volumetric flowrate, V is the tank volume (q/V signifies the residence time), E_A is the reaction activation energy, R is the universal gas constant, UA is an overall heat transfer coefficient times the tank surface area, ρ is the fluid density, C_p is the fluid heat capacity, k_0 is the rate constant, T_f is the temperature of the feed stream, C_{A0} is the inlet concentration of reactant A, ΔH_r is the heat of reaction, T is the temperature of reactor and T_c is the temperature of cooling jacket. The parameters for the CSTR in this work are detailed in Table 5.

Table 5. Reactor parameter values.

Parameter	Value
V	100 m^3
E_A/R	8750 K
$\frac{UA}{V\rho C_p}$	2.09 s^{-1}
k_0	7.2×10^{10} s^{-1}
T_f	350 K
C_{A0}	1 mol/L
$\frac{\Delta H_r}{\rho C_p}$	-209 K m^3/mol
q	100 m^3/h

A single CSTR can make multiple products by varying the concentrations of A and B in the outlet stream, which can be done by manipulating the cooling jacket temperature T_c. The cooling jacket temperature is bounded by 200 K $\leq T_c \leq$ 500 K and a constraint on movement is added as $\Delta T_c \leq 2$ K/min.

4.2. Scenarios

Five illustrative scenarios are presented to demonstrate the abilities of the noncyclic scheduling and control algorithm presented in this work, with one additional scenario introduced in Section 5 to illustrate an aspect of the iterative algorithm. The first two scenarios demonstrate the flexibility afforded by the noncyclic formulation in selecting the number of production slots and the products for production during a prediction horizon. The last three scenarios demonstrate the closed-loop performance of this formulation with process disturbances and market fluctuations. It is noted that the closed-loop performance of this formulation is also demonstrated in another work [5]; however, closed-loop selection of variable production slots is not demonstrated in another work. Additionally, analysis of computational time and comparison to a cyclic scheduling formulation are not analyzed in the companion work. All scenarios in this work use a 48 h prediction horizon, a raw material cost of $20/m^3, and a flat storage cost of $0.10/m^3/hr for all products. Slot times are given by the solutions returned by the optimization (w^*, z^*, α^*).

The first scenario as shown in Table 6 is formulated to demonstrate the flexibility of the noncyclic scheduling and control formulation. Seven products, all with large demands, are available. The noncyclic schedule is predicted to be able to select the optimal number of production slots and produce the most profitable products in the optimal production sequence rather than producing all available products in a cyclic grade wheel. The starting concentration for Scenario 1 is $C_A = 0.10$ mol/L, the steady-state concentration for Product 1.

The second scenario contains comparatively less demand per available product and is expected to produce a larger number of products than in Scenario 1. The starting concentration for Scenario is $C_A = 0.10$ mol/L, the steady-state concentration for Product 1. The third scenario is formulated with the same product demands and prices as in Scenario 2 (Table 7) but with an initial concentration of 0.22 mol/L, the steady-state concentration for Product 3. However, a process disturbance is introduced 2 h into the closed-loop simulation. The disturbance triggers a recalculation of the integrated problem.

The disturbance lasts for a duration of 1 h and the concentration rises by 0.15 mol/L. Such a disturbance could occur as a result of random equipment failure.

Table 6. Scenario 1: Product specifications.

Product	C_A (mol/L)	Max Demand (m^3)	Price ($/m^3)
1	0.10	2000	24
2	0.15	2000	29
3	0.22	2000	26
4	0.28	2000	23
5	0.34	2000	21
6	0.44	2000	21
7	0.50	2000	20

Table 7. Scenario 2: Product specifications.

Product	C_A (mol/L)	Max Demand (m^3)	Price ($/m^3)
1	0.10	1000	23
2	0.15	900	22
3	0.22	1200	29
4	0.28	860	26
5	0.34	800	25
6	0.44	1100	23
7	0.50	1400	21

The fourth scenario uses the same product specifications as Scenarios 2–3 with an initial concentration of 0.34 mol/L, but introduces a market disturbance. The updated market conditions become available 4 h into closed-loop simulations, triggering a recalculation of the scheduling and control problem. The market update includes surges in demand for Products 3–4 of 800 m^3 and 600 m^3, respectively.

The last scenario utilizes the product specifications of Scenario 1 with an initial concentration of 0.10 mol/L, but introduces market fluctuations for product prices. The updated market conditions become available 8 h into closed-loop simulations. The new prices are reflected in Table 8.

Table 8. Scenario 5: Updated market prices.

Product	C_A (mol/L)	Updated Price ($/m^3)
1	0.10	22
2	0.15	25
3	0.22	29
4	0.28	28
5	0.34	23
6	0.44	21
7	0.50	21

5. Results

The results of implementation of the scenarios and case study in Section 4 are presented in this section. Each problem is formulated in the Pyomo framework for modeling and optimization [56,57]. Nonlinear programming dynamic optimization problems are solved using orthogonal collocation on finite elements with Legrendre polynomials and Gauss–Radau roots [56–58]. The nonlinear programming dynamic optimization problems use a discretization scheme with a granularity of 1 finite element per 0.02 h of simulation time, or 50 finite elements per simulation hour with one collocation node between each finite element and are solved with the APOPT solver [59]. MILP problems are solved using the open-source COUENNE solver [60]. For each scenario, the results for closed-loop combined noncyclic scheduling and advanced control are compared to results for open-loop combined cyclic scheduling and advanced control.

5.1. Scenario 1

Scenario 1 illustrates the advantages of the flexibility of the noncyclic scheduling and control formulation. As shown in Table 6, the market demands are large for each of the seven available products. Both the cyclic and noncyclic combined scheduling and control algorithms select the most optimal production sequence, minimizing off-specification production by selecting the sequence with minimal overall transition time. However, the cyclic scheduling and control algorithm is constrained to produce each of the products on the cyclic grade wheel during the specified horizon despite the reality that only the three most profitable products should be produced. The cyclic constraint results in unnecessary transitions and off-specification production, as shown in Tables 9 and 10. Each product is inserted into the schedule even though none of the less profitable products will be produced during its turn on the cycle in order to make room for production of the most profitable products. The unnecessary transitions lead to a large decrease in profit compared to the noncyclic approach. As shown in Table 11, the noncyclic scheduling and control algorithm checks each possible number of production slots, and selects the optimal number. This minimizes unnecessary transitions and maximizes profit during the horizon considered. This scenario demonstrates the flexibility of the noncyclic combined scheduling and control approach.

Table 9. Production schedule: Scenario 1.

Formulation	Selected Production Sequence and Slot Start Times (h)						
	Slot 1	Slot 2	Slot 3	Slot 4	Slot 5	Slot 6	Slot 7
Cyclic, Traditional	P1 (0)	P2 (2.88)	P3 (23.6)	P4 (44.4)	P5 (45.1)	P7 (45.9)	P6 (47.2)
Noncyclic, Iterative	P1 (0)	P2 (6.52)	P3 (27.2)	-	-	-	-

Table 10. Economic results: Scenario 1.

Formulation	Profit ($)	Off-Spec (m³)	Manufactured Amount per Product (m³)						
			1	2	3	4	5	6	7
Cyclic, Traditional	9,984	512	288	2000	2000	0	0	0	0
Noncyclic, Iterative	18,588	148	652	2000	2000	0	0	0	0

Table 11. Alpha iterations: Scenario 1 (MILP:mixed-integer linear programming; NLP: nonlinear programming).

α	MILP	NLP
pre-iteration	-	✓ (7 Problems)
1	DF	-
2	DF	-
3	✓	-
4	✓	-
5	✓	-
6	✓	-
7	✓	-

DF: Demand Filter.

The computational requirements for the combined scheduling and control problems for Scenario 1 are outlined in Table 12. As expected, the NLP problems for both cyclic and noncyclic problems are comparable, totaling around 5 s. Due to the unique problem decomposition, the computational burden of noncyclic scheduling and control is only felt in the number of MILP iterations. The increase in the number of MILP iterations significantly raises the computational burden of the scheduling and control problem from the cyclic to the noncyclic method. However, the computational burden for both methods is reasonable for online implementation in dual-feedback loops such as those discussed in Section 2 in which an integrated scheduling and control solution is only calculated with low frequency, allowing a fast control loop to regulate the process in the absence of significant market or process disturbances and during the computation time of the integrated problem.

Table 12. Computational results: Scenario 1.

Formulation	Total Time (s)	NLP			MILP		
		#	Average (s)	Total (s)	#	Average (s)	Total (s)
Cyclic, Traditional	18.07	7	0.735	5.15	1	-	12.92
Noncyclic, Iterative	159.4	7	0.743	5.20	5	30.58	154.2

5.2. Scenario 2

In Scenario 2, the market demands for products are smaller than that of Scenario 1 and more varied. The noncyclic scheduling and control algorithm again selects the optimal number of production slots as shown in Table 13, which is again less than the number of available products on the cyclic grade wheel chosen by the cyclic scheduling and control algorithm. The integration of scheduling and control is especially evident in Scenario 2. In a scheduling formulation without integration of process dynamics and grade transition behavior, the noncyclic schedule would likely have selected the more profitable products (for example, producing product 6 rather than product 2). However, process dynamics dictate that the profit gain from producing the more profitable product (6) would be lost in the necessary grade transition and accompanying off-specification production. This scenario demonstrates the benefits of the combination of noncyclic scheduling and advanced control (see Table 14).

Computational burden in Scenario 2 is once again within the feasible range for both cyclic and noncyclic scheduling and control methods as shown in Table 15. Due to the problem decomposition, as in Scenario 1, the NLP computations require no more time for the noncyclic algorithm than the cyclic algorithm. The relative increase of the computational burden of the noncyclic scheduling and control problem is due to MILP problems in iterations. However, the demand filter shown in Table 16 is effective in reducing the number of MILP iterations required by preempting iterations for which sufficient demand is not realized.

It is important to note that the MILP problem computational burden is independent of the process model. Thus, for any given process, the relative increase in computational burden from cyclic to noncyclic scheduling and control will be independent of process model complexity. The process model only affects the nonlinear programming (or grade transition) portion of the problem decomposition, which is equally burdensome to the cyclic and noncyclic formulations.

Table 13. Production schedule: Scenario 2.

Formulation	Selected Production Sequence and Slot Start Times (h)						
	Slot 1	Slot 2	Slot 3	Slot 4	Slot 5	Slot 6	Slot 7
Cyclic, Traditional	P1 (0)	P2 (3.28)	P3 (3.96)	P4 (16.8)	P5 (26.1)	P7 (34.9)	P6 (36.2)
Noncyclic, Iterative	P1 (0)	P2 (10.0)	P3 (17.1)	P4 (29.9)	P5 (39.2)	-	-

Table 14. Economic results: Scenario 2.

Formulation	Profit ($)	Off-Spec (m^3)	Manufactured Amount per Product (m^3)						
			1	2	3	4	5	6	7
Cyclic, Traditional	3,824	512	328	0	1200	860	800	1100	0
Noncyclic, Iterative	7,420	302	1000	638	1200	860	800	0	0

Table 15. Computational results: Scenario 2.

Formulation	Total Time (s)	NLP			MILP		
		#	Average (s)	Total (s)	#	average (s)	Total (s)
Cyclic, Traditional	22.43	7	0.776	5.43	1	-	17.00
Noncyclic, Iterative	115.2	7	0.791	5.54	3	36.57	109.7

Table 16. Alpha iterations: Scenario 2.

α	MILP	B
pre-iteration	-	✓ (7 Problems)
1	DF	-
2	DF	-
3	DF	-
4	DF	-
5	✓	-
6	✓	-
7	✓	-

DF: Demand Filter.

5.3. Return Method: Additional Scenario

In Scenarios 1–2, the optimal number of slots is the first number of slots which the heuristic demand filter allows to pass to the MILP problem. This is commonly the case due to the nature of grade transitions. Fewer slots, when possible, are desirable as a grade transition is necessitated with each additional slot. It may be intuitive for the algorithm to return the solution from the first MILP problem; however, this is not always optimal, as shown through the additional scenario in Table 17. The prediction horizon duration for this additional scenario is 48 hs, and the initial concentration is 0.10 mol/L. As demonstrated in Table 18, the profit predicted by the MILP problem increases from the first slot allowed through by the heuristic demand filter upward because the difference in price between the products with high demand (Products 6–7) and those with lower demands (Products 1–5) exceeds the cost of additional grade transitions. Producing a larger number of products, with necessitated grade transitions, is more profitable than producing a smaller number of products with lower prices. This could have practical application in cases in which a selection must be made between multiple high-price, low-demand specialty products and fewer low-price, high-demand commodity products. As shown in Table 18, the most profitable number of slots is 5, though more grade transitions are necessitated. The effect of additional grade transitions exceeds the benefit from the higher price of the low-demand for combinations of more than five products, and five is selected by the algorithm as the optimal number of products to manufacture during the prediction horizon (Tables 19 and 20). More effective heuristic mechanisms for noncyclic combined scheduling and control algorithms are subjects of future work.

Table 17. Additional scenario: Product specifications.

Product	C_A (mol/L)	Max Demand (m^3)	Price ($/m^3)
1	0.10	1000	23
2	0.15	900	24
3	0.22	1200	29
4	0.28	1200	26
5	0.34	800	25
6	0.44	4000	21
7	0.50	4000	21

Table 18. Alpha Iterations: Additional scenario.

α	MILP	Predicted Profit ($)
pre-iteration	-	-
1	DF	-
2	✓	4,957
3	✓	7,077
4	✓	8,113
5	✓	12,273
6	✓	9,973
7	✓	7,601

DF: Demand Filter.

Table 19. Production schedule: Additional scenario.

Formulation	Selected Production Sequence and Slot Start Times (h)						
	Slot 1	Slot 2	Slot 3	Slot 4	Slot 5	Slot 6	Slot 7
Noncyclic, Iterative	P1 (0)	P2 (3.98)	P3 (13.7)	P4 (26.5)	P5 (39.2)	-	-

Table 20. Economic results: Additional scenario.

Formulation	Predicted Profit ($)	Off-Spec (m^3)	Manufactured Amount per Product (m^3)						
			1	2	3	4	5	6	7
Noncyclic, Iterative	12,273	302	398	900	1200	1200	800	0	0

5.4. Scenario 3

Scenario 3 demonstrates the performance of the noncyclic integrated scheduling and control method in the presence of process disturbances. The initial process and economic parameters are identical to those of Scenario 2, and the initial algorithm schedule and control choices are likewise identical. The process disturbance occurs between hours 2 and 3 of the simulation, with a concentration increase of 0.15 mol/L, from the steady-state of Product 3 (0.22 mol/L) to 0.37 mol/L. The cyclic scheduling and control shown in Table 21 is not recalculated after the disturbance. Consequently, although the process state moves to near the steady-state operating conditions of Product 5, the schedule holds and the process is driven back to Product 2 operational steady-state conditions before continuing with the cyclic grade wheel. This results in a large amount of off-specification production. The combination of unnecessary grade transitions from the cyclic grade wheel approach with the process disturbance caused a net loss through the 48 h simulation as shown in Table 22.

The noncyclic integrated scheduling and control algorithm is allowed to re-calculate the optimal scheduling and control trajectory post-disturbance. The result of the recalculation is an alteration of the schedule to produce Product 5 (0.35 mol/L), the product with the closest steady-state operating condition to the disturbed process state, post-disturbance. The recalculation cuts profit losses by reducing unnecessary grade transitions after the disturbance as well as between products which are not necessary to produce during the horizon. The noncyclic scheduling and control algorithm pulls through the simulation with a net profit, though far smaller than that of Scenario 2.

The computational burden shown in Table 23 is again within the feasible range for online dual feedback structures for initial integrated scheduling and control problems; however, the recalculation for the noncyclic algorithm requires additional computational time for nonlinear programming problems. This demonstrates that process disturbances can lead to process states or conditions from which model predictive control problems to steady-states are difficult to solve. With a dual loop feedback structure, computational requirements of larger magnitudes can be tolerable. The computational burden of the re-recalculation of the integrated scheduling and control problem for Scenario 3 is justified by the profit recovery from the recalculation. Additionally, the recalculation time can be tolerated and allowed by implanting a dual loop feedback structure in which the critical control loop is not dependent upon the solution of the integrated scheduling and control problem. Thus, the integrity of process operation (critical) is unimpeded by the computational burden of the process scheduling and control trajectory re-optimization (beneficial, but not critical). In addition, in processes in which the model complexity is too great for model predictive control approaches for integrated scheduling and control, surrogate models, time-scale bridging models, or other specifically tailored empirical or reduced order models can be substituted for the nonlinear process models in nonlinear programming optimal control and grade transition problems, potentially transforming the optimal control and grade transition problems into linear or otherwise simplified problems. Many such simplifications are demonstrated by literature reviewed in Section 2.

Table 21. Production schedule: Scenario 3.

Formulation	Selected Production Sequence and Slot Start Times (h)						
	Slot 1	Slot 2	Slot 3	Slot 4	Slot 5	Slot 6	Slot 7
Cyclic, Traditional	P3 (0)	P2 (12.0)	P1 (12.7)	P7 (16.3)	P6 (18.1)	P5 (29.9)	P4 (38.7)
Noncyclic, Iterative (Initial)	P3 (0)	P4 (12.0)	P5 (21.4)	P2 (30.1)	P1 (37.4)	-	-
Noncyclic, Iterative (Re-calc)	-	P5 (3.00)	P4 (11.6)	P3 (20.9)	P2 (31.6)	P1 (37.4)	-
Noncyclic, Iterative (Actual)	P3 (0)	Disturbance	P5 (3.00)	P4 (11.6)	P3 (20.9)	P2 (31.6)	P1 (37.4)

Table 22. Economic results: Scenario 3.

Formulation	Profit ($)	Off-Spec (m^3)	Manufactured Amount per Product (m^3)						
			1	2	3	4	5	6	7
Cyclic, Traditional (Actual)	−2096	762	291	0	988	860	800	1100	0
Noncyclic, Iterative (Actual)	4993	440	1000	500	1200	860	800	0	0

Table 23. Computational results: Scenario 3.

Formulation	Total Time (s)	NLP			MILP		
		#	Average (s)	Total (s)	#	Average (s)	Total (s)
Cyclic, Traditional	24.17	7	0.859	6.01	1	-	18.16
Noncyclic, Iterative (Initial)	112.0	7	0.836	5.05	3	35.63	106.9
Noncyclic, Iterative (Re-calc)	147.8	7	5.77	40.42	4	26.84	107.4

5.5. Scenario 4

Scenario 4 demonstrates the performance of the algorithm in the presence of volatile or changing market demand for products (see Table 24). Updated market conditions at 4 h into the simulation show demand increases for products 3 and 4 of 600 and 800 m^3, respectively. The initial economic conditions and initial process state are identical to those of Scenarios 2–3. As in Scenario 3, for comparative purposes, the cyclic schedule and control algorithm is constrained to not recalculate after the updated market conditions are made available. The result is significantly reduced profits compared to the re-optimized case. The noncyclic scheduling and control algorithm finds the initial optimal number of products to produce given the economic conditions (5), and then re-calculates the schedule when the updated market conditions are made available. The number of products to produce is reduced by one, the re-optimization to make room for the increased demand for products 2–3. This flexible re-optimization, eliminating the need for a rigid cyclic grade wheel approach, enables the scheduling and control algorithm to reduce the number of products and eliminate unnecessary grade transitions. The resulting profit increase from the noncyclic and re-optimized case is significant, as shown in Table 25.

The computational burden of the re-optimization of Scenario 4 as shown in Table 26 is far less than the burden of re-optimization in Scenario 3 as shown in Table 23. The process state for the initiation of the re-optimization was at a production steady-state, and the re-optimization computational time is comparable to that of the initial noncyclic calculation. As is the case with the other Scenarios, the computational time requirement increase for the noncyclic algorithm lies solely in the increased number of MILP problems. This relaxed computational burden increase is due to the problem decomposition described in this work.

Table 24. Production schedule: Scenario 4.

Formulation	Selected Production Sequence and Slot Start Times (h)						
	Slot 1	Slot 2	Slot 3	Slot 4	Slot 5	Slot 6	Slot 7
Cyclic, Traditional	P5 (0)	P4 (8.00)	P3 (17.3)	P2 (30.1)	P1 (30.8)	P7 (34.4)	P6 (36.2)
Noncyclic, Iterative (Initial)	P5 (0)	P4 (8.00)	P3 (17.3)	P2 (30.1)	P1 (37.4)	-	-
Noncyclic, Iterative (Re-calc)	-	P5 (4.00)	P4 (8.00)	P3 (23.3)	P2 (44.1)	-	-
Noncyclic, Iterative (Actual)	P5 (0)	P4 (8.00)	P3 (23.3)	P2 (44.1)	-	-	-

Table 25. Economic results: Scenario 4.

Formulation	Profit ($)	Off-Spec (m^3)	Manufactured Amount per Product (m^3)						
			1	2	3	4	5	6	7
Cyclic, Traditional (Actual)	2,816	546	294	0	1200	860	800	1100	0
Noncyclic, Iterative (Actual)	16,024	220	0	320	2000	1460	800	0	0

Table 26. Computational results: Scenario 4.

Formulation	Total Time (s)	NLP			MILP		
		#	Average (s)	Total (s)	#	Average (s)	Total (s)
Cyclic, Traditional	23.15	7	0.950	6.65	1	-	16.50
Noncyclic, Iterative (Initial)	122.9	7	0.964	6.75	3	38.71	116.1
Noncyclic, Iterative (Re-calc)	132.0	7	0.994	6.96	5	25.00	125.0

5.6. Scenario 5

Scenario 5 demonstrates the benefit of the noncyclic scheduling and control algorithm in the presence of volatile market prices. Updated market prices shown in Table 8 are made available at 8 h into simulation. Initial market conditions and process state are the same as in Scenario 1. Noncyclic scheduling and control initially selects three products to produce (1, 2, 3) as shown in Table 27, but responds to the price change by shifting production from products 2 and 3 to products 3 and 4, which show increased selling prices in the market update. The noncyclic scheduling and control approach demonstrates significant profit increases compared to the cyclic grade wheel approach, which is non-responsive to updated market conditions (see Table 28).

The computational burden for the noncyclic problem is again very high and infeasible for fully closed-loop regulatory control as shown in Table 29; however, as mentioned in previous discussion, the profit increase from re-optimization motivates a dual feedback structure to enable a long computation while a control feedback loop is regulating the process until the combined scheduling and control re-optimization is available. Relatively long computations are motivated by the clear potential for economic gains. Utilization of commercial solvers and commercial optimization frameworks rather than the open-source tools utilized in this work is also expected to reduce the computational burden of the formulation.

Table 27. Production schedule: Scenario 5.

Formulation	Selected Production Sequence and Slot Start Times (h)						
	Slot 1	Slot 2	Slot 3	Slot 4	Slot 5	Slot 6	Slot 7
Cyclic, Traditional	P1 (0)	P2 (2.88)	P3 (23.6)	P4 (44.4)	P5 (45.1)	P7 (45.9)	P6 (47.2)
Noncyclic, Iterative (Initial)	P1 (0)	P2 (6.52)	P3 (27.2)	-	-	-	-
Noncyclic, Iterative (Re-calc)	-	-	P3 (8.00)	P4 (28.8)	-	-	-
Noncyclic, Iterative (Actual)	P1 (0)	P2 (6.52)	P3 (8.00)	P4 (28.8)	-	-	-

Table 28. Economic results: Scenario 5.

Formulation	Profit ($)	Off-Spec (m^3)	Manufactured Amount per Product (m^3)						
			1	2	3	4	5	6	7
Cyclic, Traditional (Actual)	9,760	512	288	2000	2000	0	0	0	0
Noncyclic, Iterative (Actual)	20,820	224	652	80	2000	1844	0	0	0

Table 29. Computational results: Scenario 5.

Formulation	Total Time (s)	NLP			MILP		
		#	Average (s)	Total (s)	#	Average (s)	Total (s)
Cyclic, Traditional	19.64	7	0.827	5.79	1	-	13.85
Noncyclic, Iterative (Initial)	181.2	7	0.892	6.24	5	35.01	175.0
Noncyclic, Iterative (Re-calc)	145.3	7	0.642	4.50	6	23.47	140.8

6. Conclusions

This work demonstrates the flexibility of a noncyclic combined scheduling and advanced control framework. Economic benefit is demonstrated through a series of illustrative scenarios on a CSTR case study. Computational burden is increased compared to that of a cyclic schedule due to an increased number of MILP problems. However, economic benefit motivates the use of the noncyclic scheduling and control method in volatile market conditions and with process disturbances in a dual loop feedback structure. Economic results demonstrate the effectiveness of the flexible, noncyclic integration of scheduling and control over a cyclic grade wheel approach to combining scheduling and control when dealing with a large number of possible products within a short time-frame. The results also demonstrate the effectiveness and computational feasibility of closed-loop noncyclic scheduling and control with volatile market conditions. The benefit of a decomposition of the combined scheduling and control problem into a mixed-integer continuous-time scheduling problem and multiple separable dynamic optimization optimal control problems is demonstrated. This decomposition allows for computationally light iterations of the mixed-integer problem, enabling the noncyclic formulation within reasonable computational requirements. The benefits of shifting significant computational load, which can be calculated offline from information known a priori, is also demonstrated.

Acknowledgments: Financial support from the National Science Foundation Award 1547110 is gratefully acknowledged.

Author Contributions: This paper represents collaborative work by the authors. Damon Petersen, Logan D. R. Beal, and Derek Prestwich devised the research concepts and strategy and performed simulations. All authors were involved in the preparation of the manuscript.

Conflicts of Interest: The authors declare no conflict of interest.

Abbreviations

The following abbreviations are used in this manuscript:

NMPC	nonlinear model predictive control
MILP	mixed-integer linear programming
MIDO	mixed-integer dynamic optimization
LP	linear programming
NLP	nonlinear programming
NMPC	nonlinear model predictive control
ISC	integrated scheduling and control
CSTR	continuous-stirred tank reactor
MIDO	mixed-integer dynamic optimization
FBR	fluidized-bed reactor
SEN	state equipment network
MLDO	mixed-logic dynamic optimization
RTN	resource task network
mp-MPC	multi-parametric model predictive control
fast MPC	fast model predictive control
MINLP	mixed-integer nonlinear programming
MINFP	mixed-integer fractional programming
MILP	master scheduling problem
KPI	key performance indicator
EMPC	economic model predictive control
HIPS	high impact polystyrene
SBM	scale-bridging model
ASU	air separation unit
PWA	piecewise affine

References

1. Baldea, M.; Harjunkoski, I. Integrated production scheduling and process control: A systematic review. *Comput. Chem. Eng.* **2014**, *71*, 377–390.
2. Engell, S.; Harjunkoski, I. Optimal operation: Scheduling, advanced control and their integration. *Comput. Chem. Eng.* **2012**, *47*, 121–133.
3. Harjunkoski, I.; Nyström, R.; Horch, A. Integration of scheduling and control—Theory or practice? *Comput. Chem. Eng.* **2009**, *33*, 1909–1918.
4. Shobrys, D.E.; White, D.C. Planning, scheduling and control systems: why cannot they work together. *Comput. Chem. Eng.* **2002**, *26*, 149–160.
5. Beal, L.D.; Petersen, D.; Pila, G.; Davis, B.; Warnick, S.; Hedengren, J.D. Economic Benefit from Progressive Integration of Scheduling and Control for Continuous Chemical Processes. *Processes* **2017**, *5*, 84.
6. Capón-García, E.; Guillén-Gosálbez, G.; Espuña, A. Integrating process dynamics within batch process scheduling via mixed-integer dynamic optimization. *Chem. Eng. Sci.* **2013**, *102*, 139–150.
7. Zhuge, J.; Ierapetritou, M.G. Integration of Scheduling and Control with Closed Loop Implementation. *Ind. Eng. Chem. Res.* **2012**, *51*, 8550–8565.
8. Baldea, M.; Du, J.; Park, J.; Harjunkoski, I. Integrated production scheduling and model predictive control of continuous processes. *AIChE J.* **2015**, *61*, 4179–4190.
9. Baldea, M.; Touretzky, C.R.; Park, J.; Pattison, R.C. Handling Input Dynamics in Integrated Scheduling and Control. In Proceedings of the 2016 IEEE International Conference on Automation, Quality and Testing, Robotics (AQTR), Cluj-Napoca, Romania, 19–21 May 2016; pp. 1–6.
10. Chu, Y.; You, F. Integration of production scheduling and dynamic optimization for multi-product CSTRs: Generalized Benders decomposition coupled with global mixed-integer fractional programming. *Comput. Chem. Eng.* **2013**, *58*, 315–333.

11. Chu, Y.; You, F. Integration of scheduling and control with online closed-loop implementation: Fast computational strategy and large-scale global optimization algorithm. *Comput. Chem. Eng.* **2012**, *47*, 248–268.
12. Chu, Y.; You, F. Moving Horizon Approach of Integrating Scheduling and Control for Sequential Batch Processes. *AIChE J.* **2014**, *60*, 1654–1671.
13. Pattison, R.C.; Touretzky, C.R.; Johansson, T.; Harjunkoski, I.; Baldea, M. Optimal Process Operations in Fast-Changing Electricity Markets: Framework for Scheduling with Low-Order Dynamic Models and an Air Separation Application. *Ind. Eng. Chem. Res.* **2016**, *55*, 4562–4584.
14. Pattison, R.C.; Touretzky, C.R.; Harjunkoski, I.; Baldea, M. Moving Horizon Closed-Loop Production Scheduling Using Dynamic Process Models. *AIChE J.* **2017**, *63*, 639–651.
15. Terrazas-Moreno, S.; Flores-Tlacuahuac, A.; Grossmann, I.E. Simultaneous cyclic scheduling and optimal control of polymerization reactors. *AIChE J.* **2007**, *53*, 2301–2315.
16. Flores-Tlacuahuac, A.; Grossmann, I.E. Simultaneous Cyclic Scheduling and Control of a Multiproduct CSTR. *Ind. Eng. Chem. Res.* **2006**, *45*, 6698–6712.
17. Flores-Tlacuahuac, A.; Grossmann, I.E. Simultaneous scheduling and control of multiproduct continuous parallel lines. *Ind. Eng. Chem. Res.* **2010**, *49*, 7909–7921.
18. Terrazas-Moreno, S.; Flores-Tlacuahuac, A.; Grossmann, I.E. Lagrangean heuristic for the scheduling and control of polymerization reactors. *AIChE J.* **2008**, *54*, 163–182.
19. Zhuge, J.; Ierapetritou, M.G. An Integrated Framework for Scheduling and Control Using Fast Model Predictive Control. *AIChE J.* **2015**, *61*, 3304–3319.
20. Zhuge, J.; Ierapetritou, M.G. A Decomposition Approach for the Solution of Scheduling Including Process Dynamics of Continuous Processes. *Ind. Eng. Chem. Res.* **2016**, *55*, 1266–1280.
21. Du, J.; Park, J.; Harjunkoski, I.; Baldea, M. A time scale-bridging approach for integrating production scheduling and process control. *Comput. Chem. Eng.* **2015**, *79*, 59–69.
22. Zhuge, J.; Ierapetritou, M.G. Integration of Scheduling and Control for Batch Processes Using Multi-Parametric Model Predictive Control. *AIChE J.* **2014**, *60*, 3169–3183.
23. Beal, L.D.; Park, J.; Petersen, D.; Warnick, S.; Hedengren, J.D. Combined model predictive control and scheduling with dominant time constant compensation. *Comput. Chem. Eng.* **2017**, *104*, 271–282.
24. Beal, L.D.R.; Clark, J.D.; Anderson, M.K.; Warnick, S.; Hedengren, J.D. Combined Scheduling and Control with Diurnal Constraints and Costs Using a Discrete Time Formulation. In Proceedings of the FOCAPO (Foundations of Computer Aided Process Operations) and CPC (Chemical Process Control) 2017, Phoenix, AZ, USA, 3–12 January 2017; pp. 1–6.
25. Nishi, T.; Matsumoto, I. Petri net decomposition approach to deadlock-free and non-cyclic scheduling of dual-armed cluster tools. *IEEE Trans. Autom. Sci. Eng.* **2015**, *12*, 281–294.
26. Kim, H.J.; Lee, J.H.; Lee, T.E. Non-cyclic scheduling of a wet station. *IEEE Trans. Autom. Sci. Eng.* **2014**, *11*, 1262–1274.
27. Wikborg, U.; Lee, T.E. Noncyclic scheduling for timed discrete-event systems with application to single-armed cluster tools using pareto-optimal optimization. *IEEE Trans. Autom. Sci. Eng.* **2013**, *10*, 699–710.
28. Sakai, M.; Nishi, T. Noncyclic scheduling of dual-armed cluster tools for minimization of wafer residency time and makespan. *Adv. Mech. Eng.* **2017**, *9*, doi:10.1177/16878140176932171.
29. Kim, H.J.; Lee, J.H.; Lee, T.E. Time-Feasible Reachability Tree for Noncyclic Scheduling of Timed Petri Nets. *IEEE Trans. Autom. Sci. Eng.* **2015**, *12*, 1007–1016.
30. Kim, H.J.; Lee, J.H.; Lee, T.E. Noncyclic Scheduling of Cluster Tools With a Branch and Bound Algorithm. *IEEE Trans. Autom. Sci. Eng.* **2015**, *12*, 690–700.
31. Dias, L.S.; Ierapetritou, M.G. Integration of scheduling and control under uncertainties: Review and challenges. *Chem. Eng. Res. Des.* **2016**, *116*, 98–113.
32. Pistikopoulos, E.N.; Diangelakis, N.A. Towards the integration of process design, control and scheduling: Are we getting closer? *Comput. Chem. Eng.* **2015**, *91*, 85–92.
33. Mahadevan, R.; Doyle, F.J.; Allcock, A.C. Control-relevant scheduling of polymer grade transitions. *AIChE J.* **2002**, *48*, 1754–1764.
34. Flores-Tlacuahuac, A.; Grossmann, I.E. An effective MIDO approach for the simultaneous cyclic scheduling and control of polymer grade transition operations. *Comput. Aided Chem. Eng.* **2006**, *21*, 1221–1226.

35. Chatzidoukas, C.; Kiparissides, C.; Perkins, J.D.; Pistikopoulos, E.N. Optimal grade transition campaign scheduling in a gas-phase polyolefin FBR using mixed integer dynamic optimization. *Comput. Aided Chem. Eng.* **2003**, *15*, 744–747.
36. Chatzidoukas, C.; Perkins, J.D.; Pistikopoulos, E.N.; Kiparissides, C. Optimal grade transition and selection of closed-loop controllers in a gas-phase olefin polymerization fluidized bed reactor. *Chem. Eng. Sci.* **2003**, *58*, 3643–3658.
37. Chatzidoukas, C.; Pistikopoulos, S.; Kiparissides, C. A Hierarchical Optimization Approach to Optimal Production Scheduling in an Industrial Continuous Olefin Polymerization Reactor. *Macromol. React. Eng.* **2009**, *3*, 36–46.
38. Nie, Y.; Biegler, L.T.; Wassick, J.M. Integrated scheduling and dynamic optimization of batch processes using state equipment networks. *AIChE J.* **2012**, *58*, 3416–3432.
39. Nie, Y. Integration of Scheduling and Dynamic Optimization: Computational Strategies and Industrial Applications. Ph.D. Thesis, Carnegie Mellon University, Pittsburgh, PA, USA, July 2014.
40. Gutierrez-Limon, M.A.; Flores-Tlacuahuac, A.; Grossmann, I.E. A Multiobjective Optimization Approach for the Simultaneous Single Line Scheduling and Control of CSTRs. *Ind. Eng. Chem. Res.* **2011**, *51*, 5881–5890.
41. Gutiérrez-Limón, M.A.; Flores-Tlacuahuac, A.; Grossmann, I.E. MINLP formulation for simultaneous planning, scheduling, and control of short-period single-unit processing systems. *Ind. Eng. Chem. Res.* **2014**, *53*, 14679–14694.
42. Prata, A.; Oldenburg, J.; Kroll, A.; Marquardt, W. Integrated scheduling and dynamic optimization of grade transitions for a continuous polymerization reactor. *Comput. Chem. Eng.* **2008**, *32*, 463–476.
43. Chu, Y.; You, F. Integrated Scheduling and Dynamic Optimization of Sequential Batch Proesses with Online Implementation. *AIChE J.* **2013**, *59*, 2379–2406.
44. Chu, Y.; You, F. Integration of scheduling and dynamic optimization of batch processes under uncertainty: Two-stage stochastic programming approach and enhanced generalized benders decomposition algorithm. *Ind. Eng. Chem. Res.* **2013**, *52*, 16851–16869.
45. Chu, Y.; You, F. Integrated Scheduling and Dynamic Optimization of Complex Batch Processes with General Network Structure Using a Generalized Benders Decomposition Approach. *Ind. Eng. Chem. Res.* **2013**, *52*, 7867–7885.
46. Nyström, R.H.; Franke, R.; Harjunkoski, I.; Kroll, A. Production campaign planning including grade transition sequencing and dynamic optimization. *Comput. Chem. Eng.* **2005**, *29*, 2163–2179.
47. Nyström, R.H.; Harjunkoski, I.; Kroll, A. Production optimization for continuously operated processes with optimal operation and scheduling of multiple units. *Comput. Chem. Eng.* **2006**, *30*, 392–406.
48. Chu, Y.; You, F. Integrated scheduling and dynamic optimization by stackelberg game: Bilevel model formulation and efficient solution algorithm. *Ind. Eng. Chem. Res.* **2014**, *53*, 5564–5581.
49. Rossi, F.; Casas-Orozco, D.; Reklaitis, G.; Manenti, F.; Buzzi-Ferraris, G. A Computational Framework for Integrating Campaign Scheduling, Dynamic Optimization and Optimal Control in Multi-Unit Batch Processes. *Comput. Chem. Eng.* **2017**, *107*, 184–220.
50. Touretzky, C.R.; Baldea, M. Integrating scheduling and control for economic MPC of buildings with energy storage. *J. Process Control* **2014**, *24*, 1292–1300.
51. Terrazas-Moreno, S.; Flores-Tlacuahuac, A.; Grossmann, I.E. Simultaneous design, scheduling, and optimal control of a methyl-methacrylate continuous polymerization reactor. *AIChE J.* **2008**, *54*, 3160–3170.
52. Koller, R.W.; Ricardez-Sandoval, L.A. A Dynamic Optimization Framework for Integration of Design, Control and Scheduling of Multi-product Chemical Processes under Disturbance and Uncertainty. *Comput. Chem. Eng.* **2017**, *106*, 147–159.
53. Patil, B.P.; Maia, E.; Ricardez-Sandoval, L.A. Integration of Scheduling, Design, and Control of Multiproduct Chemical Processes Under Uncertainty. *AIChE J.* **2015**, *61*, 2456–2470.
54. Gutierrez-Limon, M.A.; Flores-Tlacuahuac, A.; Grossmann, I.E. A reactive optimization strategy for the simultaneous planning, scheduling and control of short-period continuous reactors. *Comput. Chem. Eng.* **2016**, *84*, 507–515.
55. Chu, Y.; You, F. Integrated Planning, Scheduling, and Dynamic Optimization for Batch Processes: MINLP Model Formulation and Efficient Solution Methods via Surrogate Modeling. *Ind. Eng. Chem. Res.* **2014**, *53*, 13391–13411.

56.   Hart, W.E.; Watson, J.P.; Woodruff, D.L. Pyomo: Modeling and solving mathematical programs in Python. *Math. Prog. Comput.* **2011**, *3*, 219–260.

57.   Hart, W.E.; Laird, C.; Watson, J.P.; Woodruff, D.L. *Pyomo—Optimization Modeling in Python*; Springer: Berlin/Heidelberg, Germany, 2012; Volume 67.

58.   Carey, G.; Finlayson, B.A. Orthogonal collocation on finite elements for elliptic equations. *Chem. Eng. Sci.* **1975**, *30*, 587–596.

59.   Hedengren, J.; Mojica, J.; Cole, W.; Edgar, T. APOPT: MINLP Solver for Differential Algebraic Systems with Benchmark Testing. In Proceedings of the INFORMS Annual Meeting, Pheonix, AZ, USA, 14–17 October 2012.

60.   Belotti, P.; Lee, J.; Liberti, L.; Margot, F.; Wächter, A. Branching and bounds tightening techniques for non-convex MINLP. *Optim. Methods Softw.* **2009**, *24*, 597–634.

![processes logo] *processes*

MDPI

Article

Efficient Control Discretization Based on Turnpike Theory for Dynamic Optimization

Ali M. Sahlodin [†] and Paul I. Barton *

Process Systems Engineering Laboratory, Massachusetts Institute of Technology, 77 Massachusetts Avenue, Cambridge, MA 02139, USA; sahlodin@mit.edu
* Correspondence: pib@mit.edu; Tel.: +1-617-253-6526
† Current address: Aspen Technology, 20 Crosby Dr, Bedford, MA 01730, USA.

Received: 12 November 2017; Accepted: 11 December 2017; Published: 18 December 2017

Abstract: Dynamic optimization offers a great potential for maximizing performance of continuous processes from startup to shutdown by obtaining optimal trajectories for the control variables. However, numerical procedures for dynamic optimization can become prohibitively costly upon a sufficiently fine discretization of control trajectories, especially for large-scale dynamic process models. On the other hand, a coarse discretization of control trajectories is often incapable of representing the optimal solution, thereby leading to reduced performance. In this paper, a new control discretization approach for dynamic optimization of continuous processes is proposed. It builds upon turnpike theory in optimal control and exploits the solution structure for constructing the optimal trajectories and adaptively deciding the locations of the control discretization points. As a result, the proposed approach can potentially yield the same, or even improved, optimal solution with a coarser discretization than a conventional uniform discretization approach. It is shown via case studies that using the proposed approach can reduce the cost of dynamic optimization significantly, mainly due to introducing fewer optimization variables and cheaper sensitivity calculations during integration.

Keywords: dynamic optimization; turnpike theory; control parametrization; adaptive discretization; optimal control

---

1. Introduction

The use of dynamic optimization for optimizing the performance of transient manufacturing processes has drawn much attention in recent years [1]. A typical dynamic optimization problem is to find optimal trajectories for process control variables, e.g., flowrates in a chemical process, so that a particular performance index is maximized subject to some process constraints. Numerical solution of dynamic optimization problems poses a number of difficulties. First, optimization over control trajectories gives rise to an infinite-dimensional optimization problem, which must be transformed to an approximate finite-dimensional one by discretizing control trajectories. In so-called indirect methods [2], control trajectories are discretized automatically along with the state variables as the integration routine solves a two-point boundary value problem. The resolution of the discretization depends on the error-control mechanism of the integrator. A disadvantage of indirect methods is the rather high level of expertise required to formulate the optimality conditions for problems of practical size and complexity. Moreover, it can be very difficult to solve the resulting boundary value problem, especially in the presence of constraints [2,3]. Direct methods, on the other hand, rely on a priori discretization of the control trajectories. Two main classes of direct methods are (i) the simultaneous method, where the dynamic model is also discretized, using e.g., collocation techniques and extra optimization variables [4], to arrive at a fully algebraic model and (ii) the sequential method, where the dynamic model is retained, and instead is resolved using numerical integration [5]. The focus in this work is on the sequential method that has shown capabilities for handling large-scale, stiff problems with high accuracy [6], arguably due to keeping the problem dimension relatively small [7] and the use of adaptive numerical integrators furnished with error control mechanisms.

The quality of the optimal solution is directly related to the quality of the control discretization, with a finer discretization generally offering an improved optimal solution. However, a finer discretization can lead to increased computational cost as a result of increased number of optimization variables, with too fine of a discretization also posing robustness issues [8]. For the sequential method, it also increases the cost of computing the sensitivity information by introducing extra sensitivity equations to the dynamic model. The integration of sensitivity equations can be a computationally dominant task despite the significant progress made so far regarding their efficient calculation (see [9–11]). This is especially true when the number of optimization variables is large, and can be a potentially limiting factor in applying dynamic optimization to large-scale processes.

A second difficulty with dynamic optimization arises from potentially severe nonlinearity caused by the embedded dynamic model. The nonlinearity can result in a highly nonconvex problem exhibiting many suboptimal local solutions. It is known that even small-scale dynamic systems can exhibit multiple suboptimal local solutions [12–14]. Unfortunately, a derivative-based optimization method can become trapped in any of these, which can lead to a suboptimal operation and loss of profit.

To deal with the issue of control discretization, nonuniform (i.e., not equally spaced) discretization techniques can be applied to reduce the effective number of discretization points needed. A simple way to do this is by considering the location of the discretization points as extra optimization variables. However, the extra optimization variables can cause additional nonconvexity, and thus suboptimal, local solutions, especially when the number of discretization points is fairly large. In the authors' experience, this strategy often leads to convergence difficulties, thereby defeating the point of using a coarser discretization for faster convergence. More elegant nonuniform discretization techniques have been proposed, where the discretization points over the time horizon are distributed adaptively based on the behavior of the optimal trajectories. As a result, excessive discretization can be avoided while maintaining the quality of the solution. Srinivasan et al. [15] proposed a parsimonious discretization method for optimization of batch processes with control-affine dynamics. The method relied on analysis of the information from successive batch runs for approximating the structure of the control trajectory and improving the solution in the face of fixed parametric uncertainty. Binder et al. [16] presented an adaptive discretization strategy, in which the problem with a coarse discretization of the control trajectory is first solved. Then, using a wavelet analysis of the obtained optimal solution, together with the gradient information, the discretization is refined by eliminating discretization points that are deemed unnecessary and adding new points where necessary. The problem with the refined discretization is then solved to give an improved optimal solution, and the procedure is repeated for further improvements. In a subsequent variant, Schlegel et al. [17] used a pure wavelet analysis in the refinement step in order to make the strategy better suited for problems with path constraints. In addition, Schlegel and Marquardt [18] proposed another adaptive discretization strategy that explicitly incorporates the structure of the optimal control trajectory. Specifically, the solution with a coarse discretization is used to deduce the structure based on the active or inactive status of the bound and path constraints, and the deduced structure is used to refine the discretization. A combination of the two strategies is also possible as presented in [19]. Recently, Liu et al. [20] proposed an adaptive control discretization method, in which the discretization is refined by a particular slope analysis on the approximate optimal control trajectories. The dynamic optimization problem is solved again with the refined discretization, and the procedure is repeated until a stopping criterion based on the relative improvement of the objective function is met. The foregoing strategies can be applied to a wide class of dynamic optimization problems arising from both batch and continuous processes. Nonetheless, a major drawback of them is the need for repeated solution of the dynamic optimization problem in order to arrive at a satisfactory optimal solution. Depending on the case, this may be even more costly overall than a one-time optimization with a sufficiently fine discretization.

In addition, the post-processing of results required for each refinement step would take additional time and expertise.

In this paper, a new approach to control discretization is presented. Similar to the above-mentioned strategies, it aims to create an adaptive discretization based on a deduction of the structure of the optimal trajectories. However, a different philosophy is used for this purpose. Specifically, the proposed approach is based on turnpike theory [21,22] in optimal control, which analyzes optimal control problems with respect to the structure of the optimal solution, i.e., optimal trajectories of the control and state variables. Turnpike theory has been used in the context of indirect methods in [23–25], where the structure of the optimal trajectories is exploited to approximate the resulting boundary-value problem by two initial-value problems corresponding to the initial and terminal segments of the time horizon. However, the focus in this work is on using turnpike theory for efficient control discretization in direct methods. Of particular interest is the input-state turnpike [26,27] structure, where both control and state optimal trajectories are composed of three phases: a transient phase at the beginning, followed by a non-transient phase that is close to the optimal steady state, followed by another transient phase at the end. This type of turnpike is called *steady-state* turnpike in this paper. The proposed approach exploits this structure to place the discretization points in an "optimal" way. To do so, an adaptive discretization strategy is built into the dynamic optimization formulation. In this way, the solution structure and locations of the discretization points are adjusted "dynamically" during the optimization iterations so that optimal trajectories with an adapted discretization scheme are obtained at convergence. Therefore, unlike the previous strategies, only one dynamic optimization problem is solved in the proposed approach, and no post-processing of the results is needed. Furthermore, the proposed approach helps deal with the issue of suboptimal solutions by potentially avoiding a number of such solutions whose trajectories do not conform to the turnpike structure. It is noted, however, that this approach is most suitable for optimization of transient continuous processes in which an approximate steady state may occur. It is not meant for optimization of batch or semi-batch processes, in which steady state is not possible.

The remainder of the paper is organized as follows. The problem statement along with some background on turnpike theory is presented in Section 2. The proposed control discretization approach is described in Section 3. Numerical case studies are performed in Section 4, which is followed by some conclusions in Section 5.

2. Problem Statement and Pertinent Background

The dynamic optimization problem under study can take a quite general form. For simplicity, however, a minimal formulation is considered below:

$$\min_{\mathbf{u}\in\mathcal{U}} \int_0^{t_f} J(\mathbf{x}(t,\mathbf{u}),\mathbf{u}(t))dt, \tag{1}$$

$$\text{s.t.} \quad \dot{\mathbf{x}}(t,\mathbf{u}) = \mathbf{f}(\mathbf{x}(t,\mathbf{u}),\mathbf{u}(t)), \ \forall t \in (0,t_f], \tag{2}$$

$$\mathbf{g}(\mathbf{x}(t_f,\mathbf{u}),\mathbf{u}(t_f)) \leq \mathbf{0}, \tag{3}$$

$$\mathbf{x}(0,\mathbf{u}) = \mathbf{x}_0, \tag{4}$$

where $\mathbf{u}(t) \in \mathbb{R}^{n_u}$ and $\mathbf{x}(t) \in \mathbb{R}^{n_x}$ are vectors of control and state variables, respectively, with $\mathbf{x}_0 \in \mathbb{R}^{n_x}$ the state initial conditions; $\mathcal{U}$ is the set of admissible controls defined as $\mathcal{U} \equiv \{\mathbf{u} \in (L^1([0,t_f]))^{n_u} : \mathbf{u}(t) \in U \text{ a.e. in } [0,t_f]\}$ with $U \subset \mathbb{R}^{n_u}$ bounds on $\mathbf{u}$; $\mathbf{f}$ and $\mathbf{g}$ are vector functions of appropriate sizes representing the dynamic model and process constraints, respectively. Note also that equality constraints or path constraints can be included in the formulation with no impact on the validity of the ensuing developments.

The time-dependent control variables $\mathbf{u}$ in Problem (1)–(4) give rise to an infinite-dimensional optimization problem. For a numerical solution, the problem dimension must be reduced to a finite one by discretizing $\mathbf{u}$ over the time horizon. In this work, the popular piecewise constant discretization approach is used for this purpose. With the time horizon $[0, t_f]$ discretized over N (not necessarily uniform) epochs $[0, t_1), [t_1, t_2),...,[t_{N-1}, t_N]$, $\mathbf{u}$ is approximated over each epoch by a constant parameter vector as $\mathbf{u}(t) = \mathbf{p}_k, \forall t \in [t_{k-1}, t_k)$, with $k = 1,..., N$ and $\mathbf{u}(t_f) = \mathbf{p}_N$. The discontinuities in $\mathbf{u}$ make the dynamic model (2) a continuous-discrete hybrid dynamic system [28]. The discrete behavior potentially occurs at times t_k, $k = 1,..., N-1$ (called event times). The dynamic system is said to switch from one mode to another at these times. The hybrid behavior can have implications regarding the differentiability of the dynamic system, and consequently that of the optimization problem, as discussed later.

The piecewise constant discretization reduces the optimization to one over the finite-dimensional parameter vector $\mathbf{p} = (\mathbf{p}_1, \mathbf{p}_2,..., \mathbf{p}_N) \in U^N$. However, the optimal solution of the approximated problem is generally inferior to that of the original one. To improve the solution, a finer discretization may be used by increasing the number of epochs N. The effect of increasing N, and thus the number of optimization variables, on the computational cost of the solution is twofold: (i) it increases the cost of parametric sensitivities that are calculated during solution of the dynamic model, as required by a gradient-based optimization solver; and (ii) it can increase the cost of optimization by requiring more iterations to be performed within the larger search space. In some cases, increasing N too much can also lead to robustness issues [8] and failure of the optimization solver.

The above computational concerns become even more important when dealing with large-scale models of real-life manufacturing processes. This motivates efficient control discretization strategies that can maintain the solution quality with a coarser discretization scheme. The new discretization approach presented in this paper relies on turnpike theory, which is reviewed in the following subsection.

Turnpike in Optimal Control

It appears that turnpike theory in optimal control was first discussed in the field of econometrics [22], and later gained attention in other fields including chemical processes [29]. The theory characterizes the structure of the solution of an optimal control problem by describing how the optimal control and state trajectories of a system evolve with time. It was initially investigated for optimal control problems with convex cost functions (in a minimization case). In particular, it was established that, given a time horizon $T := [0, t_f]$, the time the optimal trajectories spend outside an ϵ-neighborhood of the optimal steady state is limited to two intervals $[0, t_1]$ and $[t_f - t_2, t_f]$, where $0 \leq t_1, t_2 \leq \gamma$, with $\gamma > 0$. The system is said to be in a transient phase in these intervals. The interesting point is that γ is independent of t_f but only dependent on ϵ and the initial and final conditions of the system [30,31]. If the time horizon is long enough, i.e., $t_f > t_1 + t_2$, then a turnpike appears between t_1 and t_2, and the turnpike trajectories lie in the ϵ-neighborhood of the optimal steady state. See Figure 1 for visualization of the concept. An appealing implication of turnpike theory is that an increase in t_f will only stretch the duration of the turnpike, and has no effect on the duration and solution of the transient phases [32]. For relatively large t_f, the optimal trajectories will traverse close to the optimal steady state for most of the time horizon, and the transient phases will be short in comparison.

The extension of turnpike theory to nonconvex problems has led to generalized definitions of the turnpike, in which it may no longer be a steady state, but a time-dependent trajectory [31,33]. Nonetheless, the interest in the present work is in a steady-state turnpike for generally nonconvex problems. Recently, it was shown in [27,34] that the steady-state turnpike still occurs if the convexity assumption is replaced by a dissipativity assumption. In particular, they present a notion of strict dissipativity, and prove that, if a dynamic system is strictly dissipative with respect to a reachable optimal steady state, then the optimal trajectories will have a turnpike at that steady state. Such a turnpike emerges in practice if the time horizon is sufficiently long. The equivalence of strict dissipativity and steady-state turnpike is a key result in optimal control and has applications in stability analysis of economic model predictive control [35,36].

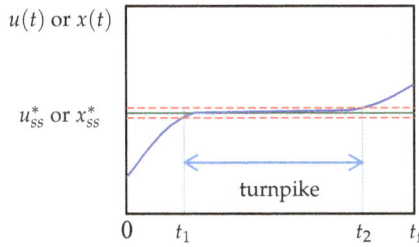

Figure 1. An optimal control or state trajectory with two transient phases and a turnpike in between. The green solid line represents the (globally) optimal steady state u_{ss}^* or x_{ss}^*, and the red dashed lines denote its ϵ-neighborhood.

3. Proposed Adaptive Control Discretization Approach

It is known that the emergence of a steady-state turnpike can be exploited in the numerical solution of optimal control problems, as noted in, e.g., [32,37], especially for problems with sufficiently long time horizons. In particular, the control trajectories need not be discretized over the entire horizon, but only over the intervals before and after the turnpike. If the turnpike interval is considerably long, this results in a coarser discretization and can reduce the computational load, as discussed earlier. However, the difficulty is that the duration of the turnpike and its location in the optimal trajectory are not known a priori. Therefore, it is not possible to adapt the discretization in advance based on the turnpike structure. To resolve this issue, this work considers embedding the tasks of approximating the turnpike structure and adaptive discretization into the dynamic optimization formulation. In this way, no pre- or post-processing for adjusting the discretization is needed. To this end, a first idea involves performing a nonuniform discretization where the duration of the epochs, i.e., $\Delta t_1 := t_1$ and $\Delta t_i := t_i - t_{i-1}$ for $i = 2, \ldots, N$ are themselves optimization variables. To account for the turnpike, the parametrized controls on one of the intermediate epochs, e.g., the middle one, can be set to their optimal steady-state values. The following optimization problem will then result:

$$\min_{\mathbf{p}_k \in U, k=1,\ldots,N, k \neq m, \Delta t_i \in [0, t_f], i=1,\ldots,N} \int_0^{t_f} J(\mathbf{x}(t, \mathbf{p}), \mathbf{p}) dt, \tag{5}$$

$$\text{s.t.} \quad \dot{\mathbf{x}}(t, \mathbf{p}) = \mathbf{f}(\mathbf{x}(t, \mathbf{p}), \mathbf{p}), \; \forall t \in (0, t_f], \tag{6}$$

$$\mathbf{g}(\mathbf{x}(t_f, \mathbf{p}), \mathbf{p}_N) \leq \mathbf{0}, \tag{7}$$

$$\sum_{i=1}^N \Delta t_i = t_f, \tag{8}$$

$$\mathbf{p}_m = \mathbf{u}_{ss}^*, \quad m = \left\lceil \frac{N}{2} \right\rceil, \tag{9}$$

$$\mathbf{x}(0, \mathbf{p}_1) = \mathbf{x}_0, \tag{10}$$

where $\lceil a \rceil$ gives the smallest integer not less than a. The optimal steady-state control $\mathbf{u}_{ss}^*$ is obtained from solving the following static optimization problem:

$$\min_{\mathbf{x}_{ss} \in X, \mathbf{u}_{ss} \in U} J(\mathbf{x}_{ss}, \mathbf{u}_{ss}), \tag{11}$$

$$\text{s.t.} \quad \mathbf{0} = \mathbf{f}(\mathbf{x}_{ss}, \mathbf{u}_{ss}), \tag{12}$$

where $X \subset \mathbb{R}^{n_x}$ are the bounds on $\mathbf{x}_{ss}$. Problem (5)–(10) will generally prescribe a nonuniform discretization to minimize the objective function. Particularly, the optimizer can adjust Δt_i so that only one epoch, here $[t_{m-1}, t_m)$, is enough to represent the steady-state turnpike. The remaining $N - 1$

epochs are automatically adjusted to cover the transient phases, where a finer discretization is needed. Problem (5)–(10) leads to a total of $n_p = (N-1) \times n_u + N$ optimization variables. The nonuniform discretization is shown schematically in Figure 2, where the middle epoch $[t_2, t_3)$ is enlarged to cover the steady-state turnpike. In case no turnpike appears in the optimal solution, the Δt_i corresponding to the turnpike can be simply pushed to zero by the optimizer. Therefore, the existence of a turnpike is not assumed and need not be verified in advance.

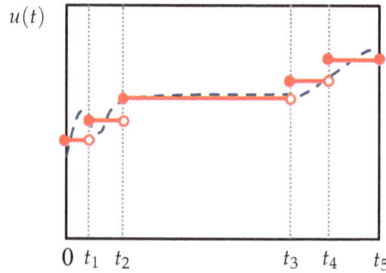

Figure 2. Schematic of an optimal control trajectory resulting from a nonuniform discretization. The dashed line illustrates the ideal optimal trajectory that might be obtained without discretization.

Note that the nonuniform discretization in Problem (5)–(10) results in a hybrid dynamic system with variable switching times t_i. For numerical reasons, a change of variable is usually used to transform Problem (5)–(10) to one with fixed switching times [38,39]. More details about this transformation are deferred to the next subsection.

With the flexibility in the values of Δt_i, it is expected that a lower N suffices to achieve the same optimal solution as in a uniform discretization. However, the addition of N optimization variables Δt_i in the nonuniform discretization strategy can adversely affect the overall computational cost and offset the benefits gained by a coarser discretization. It could contribute to more nonconvexity and thus local suboptimal solutions. The alternative proposed in this work is a *semi-uniform* adaptive discretization approach, as described in the sequel.

Semi-Uniform Adaptive Control Discretization

Here, a semi-uniform discretization approach for utilizing turnpike theory is proposed. The idea is to limit the use of nonuniform discretization where possible, while still being able to accommodate the variable durations of the steady-state turnpike and transient phases. To do so, the time horizon is split into the three phases prescribed by turnpike theory: two transient phases and a steady-state turnpike in between (see Figure 1). A uniform discretization is then applied on the transient phases, which connect to each other by one epoch representing the turnpike. The durations of these three phases are included as extra optimization variables. The approach is depicted in the left plot of Figure 3, where τ_1, τ_2, and τ_3 denote the unknown durations of the first transient, turnpike, and second transient phases, respectively. Observe that the durations of the epochs in each particular phase are equal although they are generally not equal from one phase to another, hence a semi-uniform discretization. In addition, note that each of the transient phases can take a different resolution (i.e., number of epochs). This will allow for a more accurate solution of systems where, based on experience or engineering insights, one transient phase is deemed to be considerably longer or more severe than the other, thereby requiring a finer discretization. The semi-uniform discretization approximates **u** as follows. Suppose the first transient phase (also called *startup* hereafter) is discretized into N_s uniform epochs. Then, the control vector **u** over $t \in [0, \tau_1)$ is approximated as $\mathbf{u}(t) = \mathbf{p}_k, \forall t \in [t_{k-1}, t_k)$, with $k = 1, \ldots, N_s$ and $t_{N_s} = \tau_1$. In addition, suppose the second transient (also called *shutdown* hereafter) is discretized into N_d uniform epochs. Then, the control vector **u** over $t \in [\tau_1 + \tau_2, t_f]$ is approximated as $\mathbf{u}(t) = \mathbf{p}_k, \forall t \in [t_{k-1}, t_k)$, with $k = N_s + 2, \ldots, N_s + 1 + N_d$ and $\mathbf{u}(t_{N_s+1+N_d} = t_f) = \mathbf{p}_{N_s+1+N_d}$. Notice that $\mathbf{u}(t) = \mathbf{p}_{N_s+1}$, $\forall t \in [t_{N_s}, t_{N_s+1})$ represents the one-epoch discretization corresponding to the turnpike. With this discretization scheme, the proposed approach can be formulated through the following dynamic optimization problem:

$$\min_{\mathbf{p}_k \in U, k=1,\ldots,N, k \neq N_s+1, \tau_1, \tau_2, \tau_3 \in [0, t_f]} \int_0^{t_f} J(\mathbf{x}(t, \mathbf{p}), \mathbf{p}) dt, \tag{13}$$

$$\text{s.t.} \quad \dot{\mathbf{x}}(t, \mathbf{p}) = \mathbf{f}(\mathbf{x}(t, \mathbf{p}), \mathbf{p}), \ \forall t \in (0, t_f], \tag{14}$$

$$\mathbf{g}(\mathbf{x}(t_f, \mathbf{p}), \mathbf{p}_N) \leq \mathbf{0}, \tag{15}$$

$$\sum_{i=1}^3 \tau_i = t_f, \tag{16}$$

$$\mathbf{p}_{N_s+1} = \mathbf{u}_{ss}^*, \tag{17}$$

$$\mathbf{x}(0, \mathbf{p}_1) = \mathbf{x}_0, \tag{18}$$

where $N = N_s + 1 + N_d$ is the total number of epochs. The extra variables τ_i allow for the flexibility required to approximate the transient and turnpike phases well. Unlike the nonuniform formulation, however, the proposed formulation accommodates this flexibility by introducing only three extra optimization variables, regardless of the total number of epochs considered (the total number of optimization variables is $n_p = (N-1) \times n_u + 3$ in this formulation). Furthermore, if a steady-state turnpike does not appear for a particular system, the optimizer would push τ_2 to zero. Therefore, the emergence of a turnpike need not be known a priori. However, this approach will best serve its purpose of reducing the computational cost if a turnpike exists and appears in the solution trajectories.

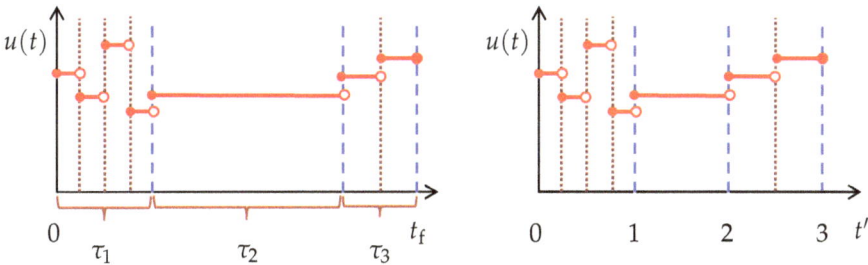

Figure 3. Schematic of the semi-uniform control discretization in the real time domain t (**left**) and the transformed time domain t' (**right**). The three transient and turnpike phases are delineated by the blue dashed lines.

Similar to Problem (5)–(10), the proposed formulation results in a hybrid dynamic system with variable switching times. This is because the locations of discretization points depend on the variables τ_i. For example, the switch for $\mathbf{u}$ from the first epoch to the second occurs at time $\frac{\tau_1}{N_s}$ according to the following statement:

$$\mathbf{u}(t) = \begin{cases} \mathbf{p}_1, & \text{if } 0 \leq t < \frac{\tau_1}{N_s}, \\ \mathbf{p}_2, & \text{if } \frac{\tau_1}{N_s} \leq t < \frac{2\tau_1}{N_s}. \end{cases} \tag{19}$$

Similarly, other switching times depend on τ_i and are not fixed. The computation of parametric sensitivities in a hybrid system where the switching times are a function of the parameters is involved and requires some assumptions for the system to ensure existence and uniqueness of the sensitivities [28]. Moreover, the parametric sensitivities are generally discontinuous over the switches [38], and additional computations are necessary to transfer the sensitivity values across the switches [28]. To avoid these numerical complications, the time transformation presented in [39,40] is used so that the switching times are fixed in the transformed formulation. In particular, the time horizon $t \in [0, t_f]$ is transformed to a new time horizon $t' \in [0, 3]$ using

$$\frac{dt}{dt'} = \tau(t'), \tag{20}$$

in which

$$\tau(t') = \begin{cases} \tau_1, & \text{if } 0 \leq t' < 1, \\ \tau_2, & \text{if } 1 \leq t' < 2, \\ \tau_3, & \text{if } 2 \leq t' \leq 3, \end{cases} \tag{21}$$

is a piecewise constant function on t'. With this transformation, Problem (13)–(18) is rewritten as:

$$\min_{\mathbf{p}_k \in U, k=1,\dots,N, k \neq N_s+1, \tau_1, \tau_2, \tau_3 \in [0, t_f]} \int_0^3 J(\mathbf{y}(t', \mathbf{p}), \mathbf{p}) dt', \tag{22}$$

$$\text{s.t.} \quad \dot{\mathbf{y}}(t', \mathbf{p}) = \tau(t') \, \mathbf{f}(\mathbf{y}(t', \mathbf{p}), \mathbf{p}), \ \forall t' \in (0, 3], \tag{23}$$

$$\mathbf{g}(\mathbf{y}(3, \mathbf{p}), \mathbf{p}_N) \leq \mathbf{0}, \tag{24}$$

$$\sum_{i=1}^3 \tau_i = t_f, \tag{25}$$

$$\mathbf{p}_{N_s+1} = \mathbf{u}_{ss}^*, \tag{26}$$

$$\mathbf{y}(0, \mathbf{p}_1) = \mathbf{y}_0, \tag{27}$$

where $\mathbf{y}(t') = \mathbf{x}(t)$. Interestingly, each of the startup, turnpike, and shutdown phases has a duration of 1 in the t' domain, as shown in the right plot of Figure 3. Moreover, the switch from one phase to another now occurs at fixed times $t' = 1$ and $t' = 2$. Accordingly, all the control switches are triggered at fixed times in the new time domain. This ensures existence and uniqueness of the parametric sensitivities and their continuity over the switches [28].

Despite the adaptive discretization strategy incorporated in Problem (22)–(27), the quality of the optimal solution can still depend on the resolutions set for the transient phases (N_s and N_d). If these resolutions are too coarse, the optimal solution may be compromised because the optimizer may have to adjust the τ_is away from their true values, e.g., in order to keep the problem feasible. Even with an adequate resolution, it is possible that the τ_is are under- or over-approximated by the local optimizer, thereby leading to an inferior solution. In some instances, it may be possible to avoid such a suboptimal solution by special modifications to the proposed formulation. In the next subsection, a variant formulation that can avoid a particular suboptimal scenario is presented.

A Variant Formulation

Figure 4 illustrates the suboptimal scenario that is dealt with by this variant formulation. The dashed lines show the optimal control trajectory that would be obtained ideally from no discretization. The left and right plots depict a suboptimal and an optimal solution, respectively, both obtained with the transient resolutions of $N_s = N_d = 2$. In the left plot, τ_2 is under-approximated. As a result, part of the actual turnpike is deemed transient by the adaptive discretization scheme. This leads to unnecessarily using three epochs for the turnpike, leaving only one epoch to represent each transient phase. Observe, however, that the suboptimal control trajectory over the turnpike is still the same as the one obtained in the optimal case. This implies that the state trajectories in part of the obtained transient phases are in fact within an ϵ-neighborhood of the optimal steady state (not shown).

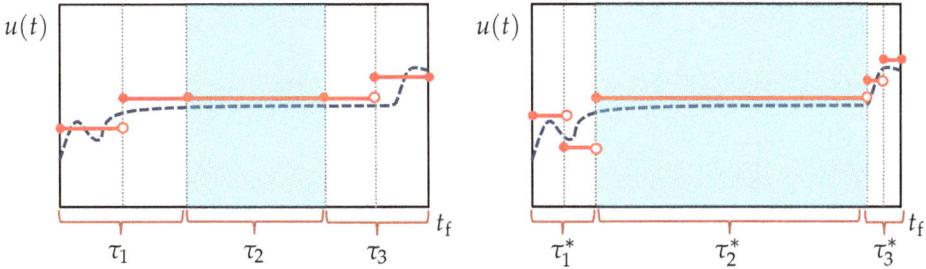

Figure 4. Adaptive semi-uniform discretization with suboptimal (**left**) and optimal (**right**) solution for τ_is. The dashed line trajectory shows the ideal optimal control as a reference.

The variant formulation avoids the above-mentioned scenario by making it infeasible to the optimization problem. Specifically, a new constraint is added to ensure that the state trajectories in the obtained transient phases are indeed outside an ϵ-neighborhood of the optimal steady state. For each state variable, this implies

$$|x(t) - x^*_{ss}| \geq \epsilon_r |x^*_{ss}| + \epsilon_a, \quad \forall t \notin [\tau_1, \tau_1 + \tau_2], \tag{28}$$

where ϵ_r and ϵ_a are relative and absolute tolerances defining the ϵ-neighborhood, respectively. Writing the squared form of Inequality (28) (to avoid potential non-differentiability) for all state variables and adding them up gives

$$\frac{1}{n_x} \sum_{i=1}^{n_x} \left(\frac{x_i(t) - x^*_{i,ss}}{\epsilon_r |x^*_{i,ss}| + \epsilon_a} \right)^2 \geq 1, \quad \forall t \notin [\tau_1, \tau_1 + \tau_2], \tag{29}$$

which, upon applying the time transformation, can be added to Problem (22)–(27) in order to rule out the possibility of the suboptimal scenario given in Figure 4. The following optimization problem will then result:

$$\min_{\mathbf{p}_k \in U, k=1,\ldots,N, k \neq N_s+1, \tau_1, \tau_2, \tau_3 \in [0, t_f]} \int_0^3 J(\mathbf{y}(t', \mathbf{p}), \mathbf{p}) dt', \tag{30}$$

$$\text{s.t.} \quad \dot{\mathbf{y}}(t', \mathbf{p}) = \tau(t') \mathbf{f}(\mathbf{y}(t', \mathbf{p}), \mathbf{p}), \quad \forall t' \in (0, 3], \tag{31}$$

$$\mathbf{g}(\mathbf{y}(3, \mathbf{p}), \mathbf{p}_N) \leq 0, \tag{32}$$

$$\frac{1}{n_x} \sum_{i=1}^{n_x} \left(\frac{x_i(t') - x^*_{i,ss}}{\epsilon_r |x^*_{i,ss}| + \epsilon_a} \right)^2 \geq 1, \quad \forall t' \notin [1, 2], \tag{33}$$

$$\sum_{i=1}^{3} \tau_i = t_f, \tag{34}$$

$$\mathbf{p}_{N_s+1} = \mathbf{u}^*_{ss}, \tag{35}$$

$$\mathbf{y}(0, \mathbf{p}_1) = \mathbf{y}_0. \tag{36}$$

Notice that (33) is a path constraint that is enforced on the transient phases only. For numerical implementation, it is converted to an equivalent terminal constraint by introducing an auxiliary state variable. See [2] for details.

4. Results and Discussion

In this section, three examples are considered to demonstrate the proposed discretization approach, and compare it against conventional uniform discretization and the nonuniform discretization approach, i.e., Problem (5)–(10), which was described earlier in this paper. The local gradient-based solver IPOPT [41] is used to solve the optimization problems. The integration of the hybrid dynamic systems and parametric sensitivity calculations are performed by the software package DAEPACK (RES Group, Needham, MA, USA) [42,43]. Note that DAEPACK is best suited for large-scale, sparse problems, and, due to the overhead associated with sparse linear algebra, may not be an optimal choice for the small-scale problems considered here. In addition, the global solver BARON [44] within the GAMS environment [45] is used to solve the static Problem (11)–(12) to global optimality. The numerical experiments are performed on a 64-bit Ubuntu 14.04 platform with a 3.2 GHz CPU.

4.1. Example 1

Consider the following dynamic optimization problem:

$$
\begin{aligned}
\min_{u} \mathcal{J}(u) := \quad & \int_0^{t_f} x_1(t,u)^2 + x_2(t,u)^2 dt \\
\text{s.t.} \quad & \dot{x}_1(t,u) = x_2(t,u) - x_1(t,u), \\
& \dot{x}_2(t,u) = -x_2(t,u) - u(t), \\
& 5x_1(t_f,u) - x_2(t_f,u)^2 = 9, \\
& (x_1(0,u), x_2(0,u)) = (0,-1), \\
& u(t) \in [-3,3], \text{ a.e. in } [0,t_f],
\end{aligned}
\tag{37}
$$

where $t_f = 20$. The optimal steady-state values required for the nonuniform discretization and proposed approaches are easily obtained by inspection as $(x_1, x_2, u) = (0,0,0)$. In addition, the tolerance values used in the variant formulation are set to $\epsilon_a = 10^{-4}$ and $\epsilon_r = 10^{-3}$. The optimal objective values and solver statistics for the uniform discretization with different numbers of epochs, nonuniform discretization, and the proposed approach are provided in Table 1. With the same number of epochs $N = 5$, the optimal solution from the proposed approach is significantly better than the one obtained from the uniform discretization. The uniform discretization was able to reach the same optimal solution only when $N = 60$ epochs were applied. In addition, with the same optimal solution $\mathcal{J}^* = 2.45$, the proposed approach converges remarkably faster, i.e., about 10 times, compared to the uniform discretization. This speed-up is due to the coarser discretization that has resulted in both fewer iterations and lower cost per iteration. The latter is because, with much fewer optimization variables, the proposed approach has a much smaller parametric sensitivity system to solve during integration. Additionally, fewer epochs mean fewer restarts of the integration at the beginning of each epoch. This can further speed up the integration. Within the proposed approach, it is seen that both the main and variant formulations yield the same solution. The variant formulation requires slightly fewer iterations to converge, which could be due to the smaller search space resulting from Constraint (33). Nevertheless, its convergence time is slightly higher; this can be attributed to the higher cost per iteration resulting from the auxiliary differential equation representing Constraint (33) and corresponding sensitivities. Finally, notice that the nonuniform discretization strategy did not converge after 1000 iterations.

Table 1. Optimal objective values, number of iterations, and CPU times for different discretization strategies in Problem (37). n_p denotes the total number of optimization variables for each strategy.

Discretization	N	n_p	$\mathcal{J}^*$	Iter	CPU (s)
Uniform	5	5	9.32	13	0.13
Uniform	60	60	2.45	103	8.02
Nonuniform	5	9	not converged	>1000	>11
Proposed	5 ($N_s = 2, N_d = 2$)	7	2.45	38	0.79
Proposed (variant)	5 ($N_s = 2, N_d = 2$)	7	2.45	33	0.89

The optimal trajectories obtained from the nonuniform discretization strategy are given in Figure 5. It is seen that the steady-state turnpike is realized for both $N = 5$ and $N = 60$ cases. However, with the coarser discretization, the turnpike is realized only partially due to the limited number of control moves available and inability to adjust their switching times optimally. Specifically, the control u must depart from its optimal steady state as early as $t = 12$ in order to use the remaining two epochs to satisfy the terminal equality constraint. On the other hand, with $N = 60$ epochs, the turnpike is realized to (possibly) its full extent due to the much higher degree of freedom in the control moves. This is to be compared with the optimal trajectories in the right plot of Figure 6, where the proposed approach is shown to be able to yield quite the same trajectories with only five control moves. The durations of the transient and turnpike phases in this case are obtained as $(\tau_1^*, \tau_2^*, \tau_3^*) = (0.6, 17.3, 2.1)$. The left plot in Figure 6 shows the optimal trajectories in the t' domain, in which the optimization problem is solved. Observe that the duration of each phase is 1 regardless of the τ_i values.

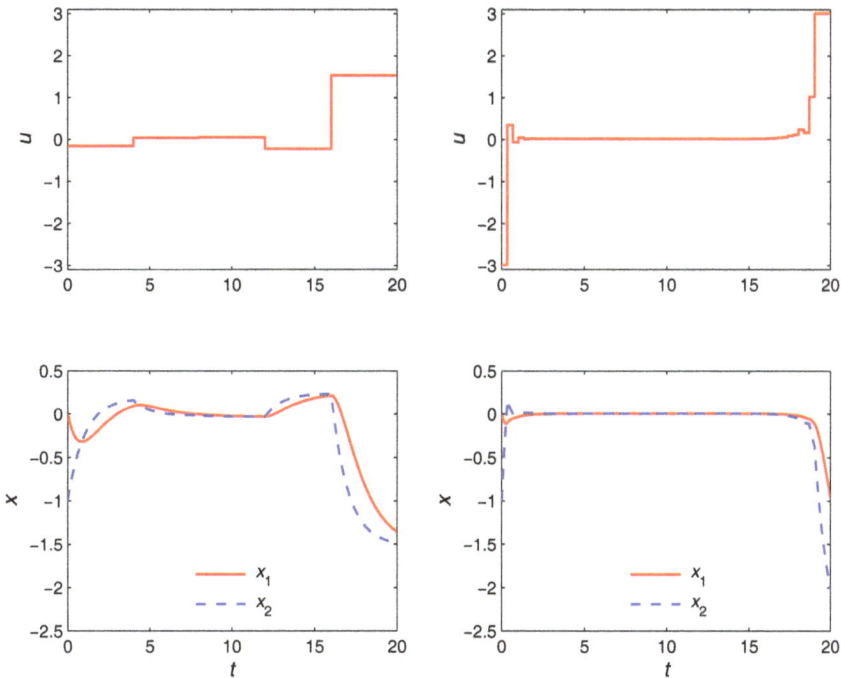

Figure 5. Optimal trajectories for Problem (37) in case of a uniform discretization with $N = 5$ (**left**) and $N = 60$ (**right**) epochs.

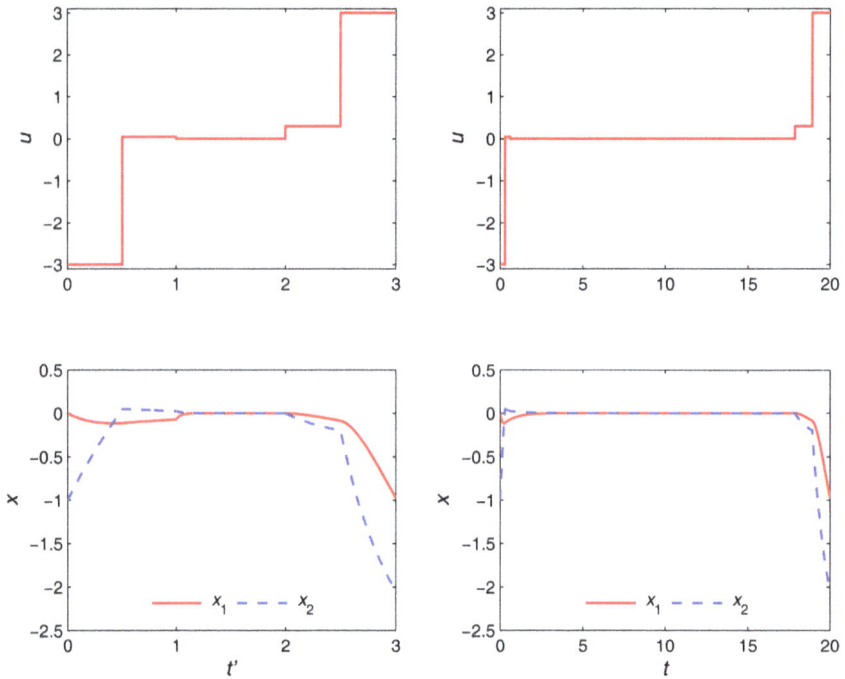

Figure 6. Optimal trajectories for Problem (37) in case of the proposed semi-uniform discretization approach with $N = 5$ epochs (both the main and variant formulations) in the t' (**left**) and t (**right**) domains.

4.2. Example 2

This example considers a non-isothermal Van de Vusse reactor adapted from [46], in which the following reactions occur:

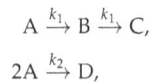

$$A \xrightarrow{k_1} B \xrightarrow{k_1} C,$$

$$2A \xrightarrow{k_2} D,$$

with B the desired product. The dynamic model of the reactor is given as:

$$
\begin{aligned}
\dot{C}_A(t) &= -k_1(T(t))C_A(t) - k_2(T(t))C_A(t)^2 + \frac{C_A^{in} - C_A(t)}{V(t)}F^{in}(t), \\
\dot{C}_B(t) &= k_1(T(t))(C_A(t) - C_B(t)) - \frac{C_B(t)F^{in}(t)}{V(t)}, \\
\dot{C}_C(t) &= k_1(T(t))C_B(t) - \frac{C_C(t)F^{in}(t)}{V(t)}, \\
\dot{C}_D(t) &= \tfrac{1}{2}k_2(T(t))C_A(t)^2 - \frac{C_D(t)F^{in}(t)}{V(t)}, \\
\dot{T}(t) &= h(t) + \alpha(T_c(t) - T(t)) + F^{in}(t)\frac{T^{in} - T(t)}{V(t)}, \\
\dot{T}_c(t) &= \beta(T(t) - T_c(t)) - \gamma P_c(t), \\
\dot{V}(t) &= F^{in}(t) - F^{out}(t), \\
F^{out}(t) &\equiv \eta\sqrt{V(t)}, \\
h(t) &\equiv -\delta(k_1(T(t))(C_A(t)\Delta H_{AB} + C_B(t)\Delta H_{BC}) + k_2(T(t))C_A(t)^2\Delta H_{AD}), \\
k_1(T(t)) &\equiv k_{0,1}\exp\left(\frac{-E_1}{T(t)+273.15}\right), \\
k_2(T(t)) &\equiv k_{0,2}\exp\left(\frac{-E_2}{T(t)+273.15}\right),
\end{aligned}
\tag{38}
$$

where C_i denotes the concentration of i; T and T_c are the temperatures of the reactor and cooling jacket, respectively; V is the reaction volume; and F^{out} is the outlet flowrate. Here, the original model in [46] has been modified to allow a varying volume. It is worth noting that, for the original model, conditions ensuring strict dissipativity, and, thus, potential emergence of turnpike have been verified in [27]. Nonetheless, such a result is not required in this work since it makes no assumption on the presence of a turnpike. The initial conditions are given as:

$$(C_A(0), C_B(0), C_C(0), C_D(0)) = (0, 0, 0, 0) \text{ mol m}^{-3},$$
$$(T(0), T_c(0)) = (108, 107.7)\,^\circ C, \ V(0) = 10^{-3}\,\text{m}^3. \tag{39}$$

In addition, the model constants are provided in Table 2. For simplicity, the normalized heat transfer coefficients α and β are assumed to be independent of the volume V [47]. The operation takes $t_f = 10$ h. As a safety precaution, the reactor temperature T must not exceed $110\,^\circ C$ anytime during the operation. Similarly, the concentration of D must not exceed 500 mol m^{-3}. At the final time, the reactor volume V must be 0.01 m^3 or less. The optimization variables are the inlet flowrate F^{in} and the cooling power P_c, which are allowed to vary within $[0, 40]$ m^3 h^{-1} and $[0, 4000]$ kJ h^{-1}, respectively. With the goal of maximizing the production of B, the dynamic optimization problem can be formulated as:

$$\min_{F^{in}, P_c} \mathcal{J}(F^{in}, P_c) := - \int_0^{t_f} F^{out}(t, F^{in}) C_B(t, [F^{in}, P_c]) dt,$$

$$\begin{aligned}
\text{s.t.} \quad & \text{Dynamic model (38),} \quad \forall t \in (0, t_f], \\
& V(t_f, F^{in}) - 10^{-2} \le 0, \\
& T(t, [F^{in}, P_c]) \le 110, \quad && \forall t \in [0, t_f], \\
& C_D(t, [F^{in}, P_c]) \le 500, \quad && \forall t \in [0, t_f], \\
& \text{Initial conditions (39),} \\
& F^{in}(t) \in [0, 40], \ P_c(t) \in [0, 4000], \ \text{a.e. in } [0, t_f].
\end{aligned} \tag{40}$$

The optimal steady-state values required for the nonuniform discretization and proposed approaches are obtained from

$$\min_{F^{in}_{ss}, P_{c,ss}, \mathbf{C}_{ss}, V_{ss}, T_{ss}, T_{c,ss}} \quad J := -F^{out}_{ss} C_{B,ss},$$

$$\begin{aligned}
\text{s.t.} \quad & \text{steady-state model,} \\
& F^{in}_{ss} \in [0, 40], \quad P_{c,ss} \in [0, 4000], \\
& C_{A,ss} \in [0, 5000], \quad C_{B,ss} \in [0, 2000], \\
& C_{C,ss} \in [0, 2000], \quad C_{D,ss} \in [30, 500], \\
& V_{ss} \in [0.01, 5], \quad T_{ss} \in [0, 110], \ T_{c,ss} \in [0, 150],
\end{aligned} \tag{41}$$

where $\mathbf{C}$ is the vector containing all the concentrations, and the steady-state model refers to the dynamic model (38) with the time derivatives set to zero. Notice that the path constraints on C_D and T are included in Problem (41) as upper bounds for the corresponding variables. Except these and the lower bound of zero on the concentrations, other bounds on the state variables have no process implications and are placed so that the global solver can proceed. Once a global optimum is found, the solution must be checked to make sure these arbitrary bounds are not active.

The global solution of Problem (41) is obtained in less than 10^{-3} s, with the optimum point $F^{in,*}_{ss} = 40$ m^3 h^{-1}, $P^*_{c,ss} = 3040.6$ kJ h^{-1}, $(C^*_{A,ss}, C^*_{B,ss}, C^*_{C,ss}, C^*_{D,ss}) = (2944.7, 977.9, 486.2, 345.6)$ mol m^{-3}, $(T^*_{ss}, T^*_{c,ss}) = (110, 106.5)\,^\circ C$, and $V^*_{ss} = 1.8$ m^3.

Table 2. Constants for the Van de Vusse reactor model (38).

$k_{0,1}$	1.29×10^{12}	h^{-1}	α	30.828	h^{-1}
$k_{0,2}$	9.04×10^{6}	$m^3 \, (\text{mol h})^{-1}$	β	86.688	h^{-1}
E_1	9758.3	K	γ	0.1	$K \, kJ^{-1}$
E_2	8560	K	δ	3.52×10^{-4}	$m^3 \, K \, kJ^{-1}$
ΔH_{AB}	4.2	$kJ \, mol^{-1}$	η	30	$m^{3/2} \, h^{-1}$
ΔH_{BC}	-11	$kJ \, mol^{-1}$	T_{in}	104.9	$^\circ C$
ΔH_{AD}	-41.85	$kJ \, mol^{-1}$	C_A^{in}	5.10×10^{3}	$mol \, m^{-3}$

The optimal solution and solver statistics for the different discretization strategies are given in Table 3. For the variant formulation of the proposed approach, the settings $\epsilon_a = 10^{-4}$ and $\epsilon_r = 10^{-1}$ are used. Similar to the previous example, the proposed approach yields a better optimal solution than the uniform discretization with the same number of epochs. The optimal solution of the latter improves by increasing the number of epochs to $N = 60$. Nonetheless, it is still inferior to the one obtained from the proposed approach with $N = 7$. Increasing N to 70 did not lead to an improved result as the solver failed after 207 iterations. The same problem occurred with $N = 80$. This shows that a finer discretization is not always beneficial in practice as it can lead to numerical issues. Similarly, the nonuniform discretization with $N = 7$ terminated with a failure message indicating local infeasibility. However, this problem is known to be feasible because a more constrained version of it, i.e., the proposed semi-uniform discretization approach with $N = 7$, is feasible (see Table 3). Therefore, the reported local infeasibility is only due to numerical issues that apparently arise from including the duration of epochs as extra optimization variables. In terms of the solution speed, the CPU times show that the proposed approach (main formulation) converges to a *better* solution about 10 times faster than the uniform discretization with $N = 60$. This is in spite of the significantly more iterations that are taken by the proposed approach. The speed-up is even more remarkable for the variant formulation, i.e., about 37-fold.

Table 3. Optimal objective values, number of iterations, and CPU times for different discretization strategies in Problem (40). n_p denotes the total number of optimization variables for each strategy.

Discretization	N	n_p	$\mathcal{J}^*$	Iter	CPU (s)
Uniform	7	14	-3.34×10^5	251	14.7
Uniform	60	120	-3.84×10^5	227	798
Uniform	70	140	failed	207	645
Uniform	80	160	failed	295	1160
Nonuniform	7	19	failed	1212	134
Proposed	$7 \, (N_s = 3, N_d = 3)$	15	-3.87×10^5	800	75.2
Proposed (variant)	$7 \, (N_s = 3, N_d = 3)$	15	-3.87×10^5	210	21.4

The optimal trajectories for control variables and a selection of the state variables in the case of the uniform discretization with $N = 7, 60$ and the proposed approach (both variants) are plotted in Figures 7 and 8, respectively. The units for the quantities are consistent with those reported in Table 2. Similar observations as in the previous example hold here, and are omitted for brevity. The only notable additional point is that, here, the optimal trajectories from the main and the variant formulations are not exactly the same, although the difference can hardly be noticed. Moreover, the main and the variant formulations converge to slightly different values for the triplet $(\tau_1^*, \tau_2^*, \tau_3^*)$, i.e., $(5.4, 93.2, 1.4)$ and $(2.9, 95.7, 1.4)$ h for the main and variant formulations, respectively. The optimal objective value from both formulations, however, is almost the same despite the slight difference in the computed optima.

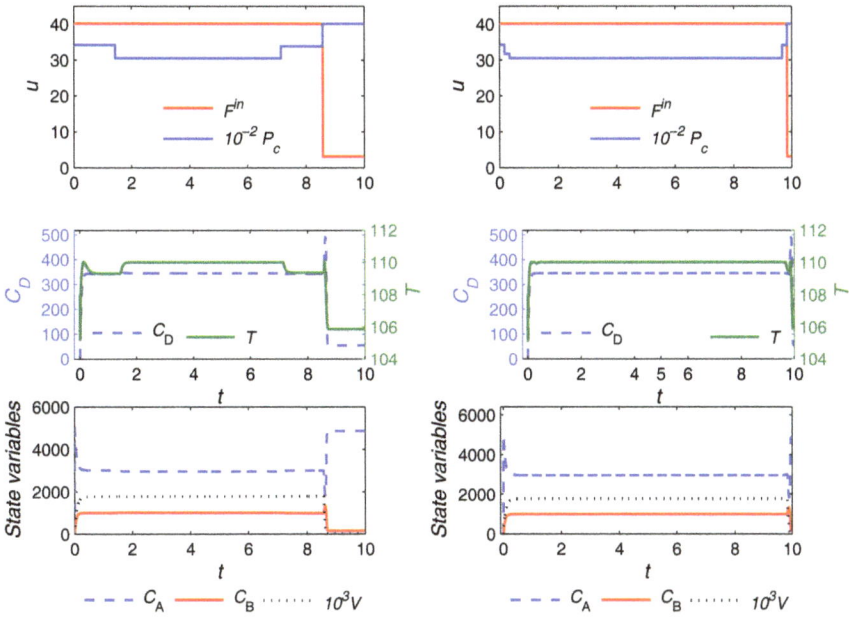

Figure 7. Optimal trajectories for Problem (40) in the case of a uniform discretization with $N = 7$ (**left**) and $N = 60$ (**right**) epochs.

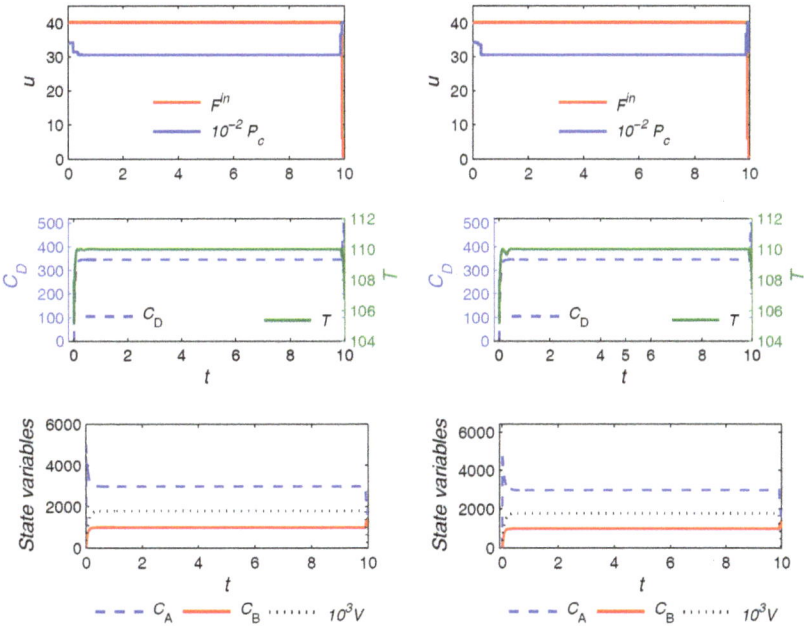

Figure 8. Optimal trajectories for Problem (40) in the case of the main (**left**) and the variant (**right**) formulations of the proposed approach with $N = 7$ epochs.

Shorter Time Horizon

Here, the Van de Vusse reactor is revisited with a time horizon that is too short for the turnpike to appear. The purpose of this case study is to show that the proposed discretization approach still works in such conditions, although it may not be computationally more efficient than conventional discretization. To this end, the final time is reduced to $t_f = 0.2$ h, and the problem is solved using the uniform discretization with $N = 5, 7$ and the proposed approach with $N = 5$ ($N_s = 2, N_d = 2$). The computational results are given in Table 4, and the optimal trajectories for the case of $N = 5$ are plotted in Figures 9 and 10. It is seen that the optimal trajectories do not reach a steady-state turnpike. Despite this, the proposed approach is still able to solve the problem, and even converge to a slightly better solution than the uniform discretization with $N = 5$. The durations of the transient and turnpike phases are obtained as $(\tau_1^*, \tau_2^*, \tau_3^*) = (0.11, 0, 0.09)$ h. Interestingly enough, it is seen that the duration of the turnpike is pushed to zero by the optimizer, reducing the effective number of epochs to four. The convergence to a better solution despite the absence of a turnpike and lower number of epochs can be explained by the inherent flexibility of the proposed approach in adjusting the location and duration of epochs. Note, however, that the uniform discretization with $N = 7$ is able to yield yet an improved solution. In both cases, the solution with the uniform discretization converges somewhat faster, than the proposed approach. This suggests that, for very short time horizons, the extra computational overhead introduced by the proposed approach may not be offset by the reduction in the number of epochs.

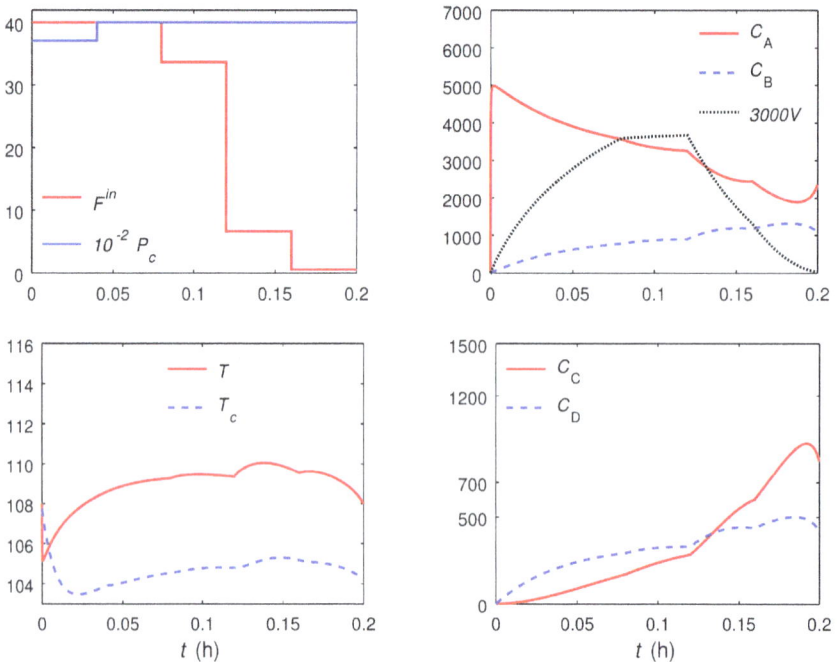

Figure 9. Optimal trajectories for Problem (40) with a short time horizon solved using a uniform discretization and $N = 5$ epochs.

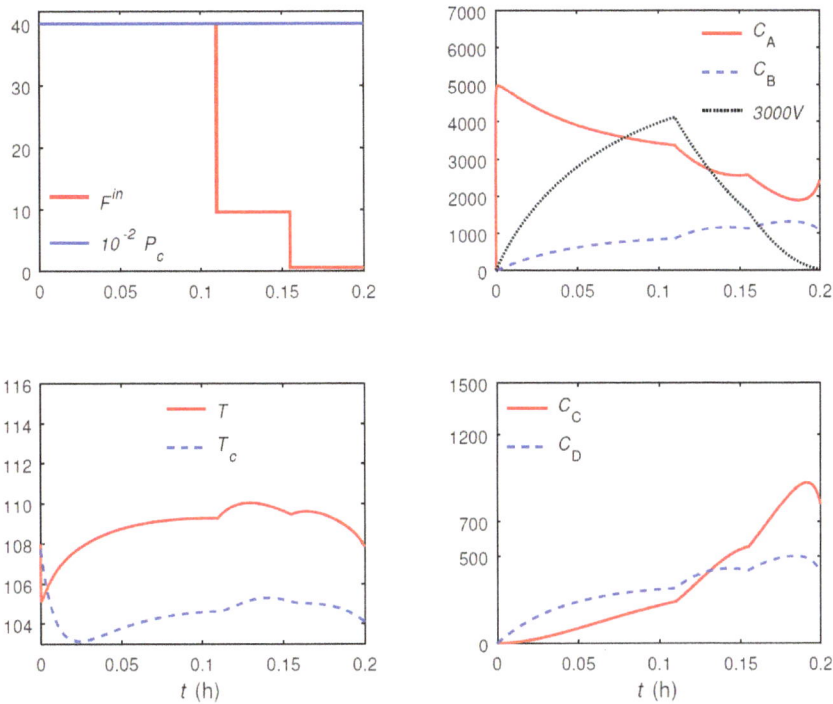

Figure 10. Optimal trajectories for Problem (40) with a short time horizon solved using the proposed approach and $N = 5$ epochs.

Table 4. Optimal objective values, number of iterations, and CPU times for different discretization strategies in Problem (40) for the case of a short time horizon.

Discretization	N	n_p	$\mathfrak{J}^*$	Iter	CPU (s)
Uniform	5	10	3987	44	1.65
Uniform	7	14	4058	53	2.29
Proposed	5 ($N_s = 2, N_d = 2$)	11	4009	69	2.77
Proposed (variant)	5 ($N_s = 2, N_d = 2$)	11	4009	53	2.33

4.3. Example 3

Finally, dynamic optimization of a continuous stirred-tank reactor adapted from [47] is presented. The following reactions take place in the reactor:

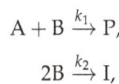

$$A + B \xrightarrow{k_1} P,$$

$$2B \xrightarrow{k_2} I,$$

where P and I are the desired and undesired products, respectively; k_1 and k_2 are the kinetic constants. Two pure streams at the flowrates F_A^{in} and F_B^{in} and concentrations C_A^{in} and C_B^{in}, respectively, enter the reactor. The reactor is modeled as:

$$
\begin{aligned}
\dot{C}_A(t) &= -k_1 C_A(t) C_B(t) + \tfrac{F_A^{in}(t)}{V(t)}(C_A^{in} - C_A(t)) - \tfrac{F_B^{in}(t)}{V(t)} C_A(t), \\
\dot{C}_B(t) &= -k_1 C_A(t) C_B(t) - 2k_2 C_B^2(t) + \tfrac{F_B^{in}(t)}{V(t)}(C_B^{in} - C_B(t)) - \tfrac{F_A^{in}(t)}{V(t)} C_B(t), \\
\dot{C}_P(t) &= k_1 C_A(t) C_B(t) - \tfrac{(F_A^{in}(t)+F_B^{in}(t))}{V(t)} C_P(t), \\
\dot{C}_I(t) &= k_2 C_B^2(t) - \tfrac{(F_A^{in}(t)+F_B^{in}(t))}{V(t)} C_I(t), \\
\dot{V}(t) &= F_A^{in}(t) + F_B^{in}(t) - F^{out}(t),
\end{aligned}
\tag{42}
$$

with $F^{out}(t) \equiv \alpha \sqrt{V(t)}$ being the outlet flow rate, where α is a positive constant, and the initial conditions are given by

$$
C_A(0) = C_{A0}, C_B(0) = C_{B0}, C_P(0) = C_{P0}, C_I(0) = C_{I0}, V(0) = V_0.
\tag{43}
$$

The model parameters and the initial conditions are found in Table 5. The operation takes $t_f = 50$ min, and the objective is to maximize production of P during this period. The inlet flow rates are taken as the optimization variables. The following dynamic optimization problem is formulated:

$$
\min_{F_A^{in}, F_B^{in}} \mathcal{J}(F_A^{in}, F_B^{in}) := - \int_0^{t_f} F^{out}(t, [F_A^{in}, F_B^{in}]) C_P(t, [F_A^{in}, F_B^{in})]) dt,
$$

$$
\begin{aligned}
\text{s.t.} \quad & \text{Dynamic model (42),} \quad \forall t \in (0, t_f], \\
& V(t_f, [F_A^{in}, F_B^{in}]) - 10^{-3} \leq 0, \\
& C_I(t, [F_A^{in}, F_B^{in}]) \leq 0.14, \qquad \forall t \in [0, t_f], \\
& \text{Initial conditions (43),} \\
& F_A^{in} \in [0, 0.01], F_B^{in} \in [0.002, 0.01], \text{ a.e. in } [0, t_f].
\end{aligned}
\tag{44}
$$

Table 5. Constants and initial conditions for the model (42).

	Parameters			Initial Conditions	
k_1	0.8	L (mol min)$^{-1}$	C_{A0}	0	mol L^{-1}
k_2	0.5	L (mol min)$^{-1}$	C_{B0}	0	mol L^{-1}
C_A^{in}	5	mol L^{-1}	C_{P0}	0	mol L^{-1}
C_B^{in}	3	mol L^{-1}	C_{I0}	0	mol L^{-1}
α	0.119	L$^{1/2}$ min^{-1}	V_0	0.001	L

The optimal steady-state values required for the nonuniform discretization and proposed approaches are obtained from

$$
\min_{F_{A,ss}^{in}, F_{B,ss}^{in}, C_{ss}, V_{ss}} \quad J := -F_{ss}^{out} C_{P,ss},
$$

$$
\begin{aligned}
\text{s.t.} \quad & \text{steady-state model,} \\
& F_{A,ss}^{in} \in [0, 0.01], \quad F_{B,ss}^{in} \in [0.002, 0.01], \\
& C_{A,ss} \in [0, 2], \quad C_{B,ss} \in [0, 2], \\
& C_{P,ss} \in [0, 2], \quad C_{I,ss} \in [0, 0.14], \\
& V_{ss} \in [0, 1],
\end{aligned}
\tag{45}
$$

where **C** is the vector containing all the concentrations, and the steady-state model refers to Problem (42) with the time derivatives set to zero. Similar to the previous example, bounds on the volume and the concentrations of A, B, and P have no process implications and are placed so that the global solver can proceed. The obtained global solution is accepted if these arbitrary bounds are not active.

The global solution of Problem (45) is obtained in 0.016 s, with the optimum point $F_{A,ss}^{in,*} = F_{B,ss}^{in,*} = 0.01$ L min^{-1}, $(C_{A,ss}^*, C_{B,ss}^*, C_{P,ss}^*, C_{I,ss}^*) = (1.69, 0.43, 0.82, 0.13)$ mol L^{-1}, and $V_{ss}^* = 0.03$ L.

Table 6 shows the optimal solution and solver statistics for the uniform, nonuniform, and proposed discretization strategies. The tolerances $\epsilon_a = 0$ and $\epsilon_r = 10^{-2}$ are used for the variant formulation of the proposed approach. With $N = 5$ epochs, the proposed strategy and its variant yield a considerably improved solution compared to the uniform discretization. The variant formulation outperforms the main one in both number of iterations and CPU time. It also slightly outperforms the uniform discretization with $N = 5$ in the same respects. The durations of the transient and turnpike phases for the proposed approach (both formulations) are obtained as $(\tau_1^*, \tau_2^*, \tau_3^*) = (0, 45.3, 4.7)$ min.

Table 6. Optimal objective values, number of iterations, and CPU times for different discretization strategies in Problem (44). n_p denotes the total number of optimization variables for each strategy.

Discretization	N	n_p	$\mathcal{J}^*$	Iter	CPU (s)
Uniform	5	10	0.66	31	1.3
Uniform	14	28	0.734	23	1.26
Uniform	20	40	0.739	27	2.38
Uniform	21	42	0.741	34	3.53
Nonuniform	2	4	0.731	35	0.59
Nonuniform	3	7	0.742	145	3.28
Proposed	5 ($N_s = 2, N_d = 2$)	11	0.741	29	1.83
Proposed (variant)	5 ($N_s = 2, N_d = 2$)	11	0.741	24	1.18

The uniform strategy converges to the same solution with a much higher number of epochs ($N = 21$) and a markedly higher computational time (93% and 200% slower compared to the main and variant formulations, respectively). Here, the nonuniform strategy converges quite closely to the optimal solution of $\mathcal{J}^* = 0.741$ with only two epochs. With only one more epoch, the nonuniform strategy is able to reach this solution (and slightly improves upon it). Nonetheless, the solution takes a much higher CPU time than the one with the proposed strategy, i.e, 1.8 and 2.8 times higher than the main and variant formulations, respectively. The durations of the three epochs used by the nonuniform discretization are obtained as $(45.5, 0.1, 4.4)$ min.

The optimal trajectories are shown in Figures 11 and 12. The trajectories for the main and variant formulations of the proposed approach are identical.

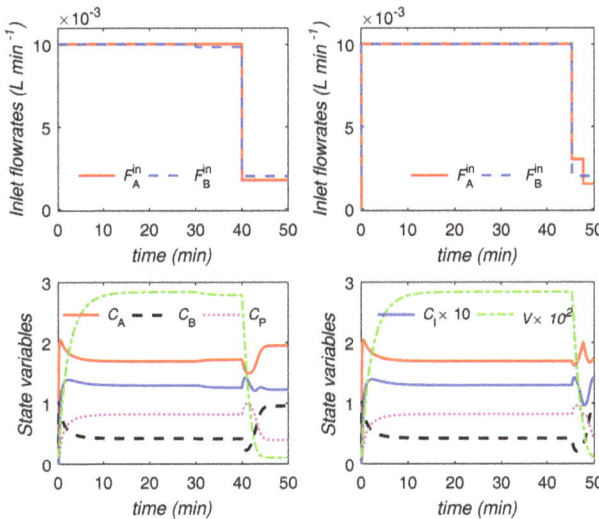

Figure 11. Optimal trajectories for Problem (44) in the case of the uniform (**left**) and proposed (**right**) discretization methods both with $N = 5$ epochs.

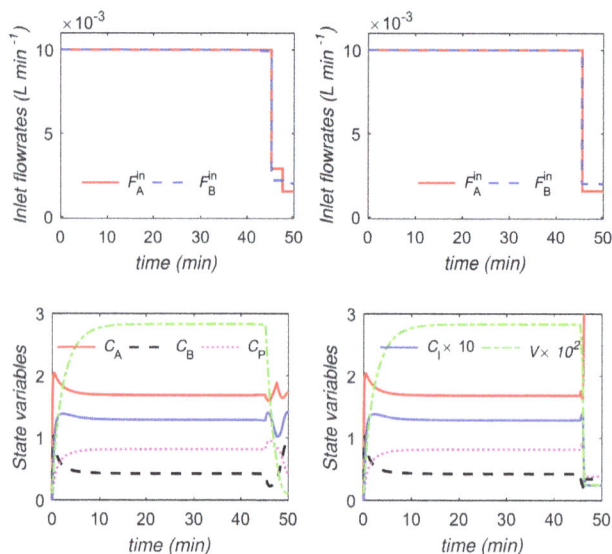

Figure 12. Optimal trajectories for Problem (44) in the case of the uniform discretization with $N = 21$ (**left**) and the nonuniform discretization with $N = 3$ (**right**) epochs.

5. Conclusions

A new adaptive control discretization approach for efficient dynamic optimization is proposed. The approach is based on turnpike theory in optimal control and is most suitable for continuous systems with sufficiently long time horizons during which steady state is likely to emerge. However, it can also be applied to other systems with a steady-state solution, whether or not the steady state will actually appear in the solution trajectories. The special semi-uniform discretization enables approximating the turnpike structure with a minimal number of epochs, while avoiding the robustness issues that can be encountered with a full nonuniform discretization strategy. Unlike some other adaptive discretization techniques, the proposed adaptive discretization is built directly into the problem formulation. Thus, one would need to solve only one optimization problem instead of a series of successively refined problems. Another advantage of the proposed approach is the use of globally optimal steady-state values in the formulation that helps the optimizer avoid suboptimal solutions in case a steady-state turnpike emerges. It is shown that the proposed approach, especially the variant formulation, can significantly reduce the computational cost of dynamic optimization for systems of interest. However, a downside of the variant formulation is that the tolerance values can impact the performance of the numerical solution, and finding appropriate values for them may not be trivial. In this case, one may choose to use the main formulation instead, as it is adequately superior and requires no tuning parameters. Future work may include applying the proposed approach to large-scale processes and optimal campaign continuous manufacturing problems, as described in [47].

Acknowledgments: Financial support from the Novartis-MIT Center for Continuous Manufacturing (Cambridge, MA, USA) is gratefully acknowledged.

Author Contributions: This work was done under the supervision of Paul I. Barton. Both authors contributed to the development of the proposed approach. Ali M. Sahlodin also performed the case studies and prepared the manuscript.

Conflicts of Interest: The authors declare no conflict of interest.

References

1. Cervantes, A.; Biegler, L. Optimization Strategies for Dynamic Systems. In *Encyclopedia of Optimization*; Floudas, C.A., Pardalos, P.M., Eds.; Springer: Boston, MA, USA, 2009; pp. 2847–2858.

2. Chachuat, B. *Nonlinear and Dynamic Optimization: From Theory to Practice-IC-32: Spring Term 2009*; Polycopiés de l'EPFL, EPFL: Lausanne, Switzerland, 2009.

3. Von Stryk, O.; Bulirsch, R. Direct and indirect methods for trajectory optimization. *Ann. Oper. Res.* **1992**, *37*, 357–373.

4. Biegler, L. An overview of simultaneous strategies for dynamic optimization. *Chem. Eng. Process.* **2007**, *46*, 1043–1053.

5. Kraft, D. On Converting Optimal Control Problems into Nonlinear Programming Problems. In *Computational Mathematical Programming*; Schittkowski, K., Ed.; Springer: Berlin/Heidelberg, Germany, 1985; Volume 15, pp. 261–280.

6. Binder, T.; Blank, L.; Bock, H.; Bulirsch, R.; Dahmen, W.; Diehl, M.; Kronseder, T.; Marquardt, W.; Schlöder, J.; von Stryk, O. Introduction to Model Based Optimization of Chemical Processes on Moving Horizons. In *Online Optimization of Large Scale Systems*; Grötschel, M., Krumke, S., Rambau, J., Eds.; Springer: Berlin/Heidelberg, Germany, 2001; pp. 295–339.

7. Hartwich, A.; Marquardt, W. Dynamic optimization of the load change of a large-scale chemical plant by adaptive single shooting. *Comput. Chem. Eng.* **2010**, *34*, 1873–1889.

8. Binder, T.; Blank, L.; Dahmen, W.; Marquardt, W. Grid refinement in multiscale dynamic optimization. In *European Symposium on Computer Aided Process Engineering-10*; Pierucci, S., Ed.; Elsevier: Amsterdam, The Netherlands, 2000; Volume 8, pp. 31–36.

9. Feehery, W.F.; Tolsma, J.E.; Barton, P.I. Efficient sensitivity analysis of large-scale differential-algebraic systems. *Appl. Numer. Math.* **1997**, *25*, 41–54.

10. Özyurt, D.B.; Barton, P.I. Cheap Second Order Directional Derivatives of Stiff ODE Embedded Functionals. *SIAM J. Sci. Comput.* **2005**, *26*, 1725–1743.

11. Özyurt, D.B.; Barton, P.I. Large-Scale Dynamic Optimization Using the Directional Second-Order Adjoint Method. *Ind. Eng. Chem. Res.* **2005**, *44*, 1804–1811, doi:10.1021/ie0494061.

12. Luus, R.; Cormack, D.E. Multiplicity of solutions resulting from the use of variational methods in optimal control problems. *Can. J. Chem. Eng.* **1972**, *50*, 309–311.

13. Luus, R.; Dittrich, J.; Keil, F.J. Multiplicity of solutions in the optimization of a bifunctional catalyst blend in a tubular reactor. *Can. J. Chem. Eng.* **1992**, *70*, 780–785.

14. Banga, J.R.; Seider, W.D. Global optimization of chemical processes using stochastic algorithms. In *State of the Art in Global Optimization: Computational Methods and Applications*; Floudas, C., Pardalos, P., Eds.; Kluwer Academic Publishers: Dordrecht, The Netherlands, 1996; pp. 563–583.

15. Srinivasan, B.; Primus, C.; Bonvin, D.; Ricker, N. Run-to-run optimization via control of generalized constraints. *Control Eng. Pract.* **2001**, *9*, 911–919.

16. Binder, T.; Cruse, A.; Villar, C.C.; Marquardt, W. Dynamic optimization using a wavelet based adaptive control vector parameterization strategy. *Comput. Chem. Eng.* **2000**, *24*, 1201–1207.

17. Schlegel, M.; Stockmann, K.; Binder, T.; Marquardt, W. Dynamic optimization using adaptive control vector parameterization. *Comput. Chem. Eng.* **2005**, *29*, 1731–1751.

18. Schlegel, M.; Marquardt, W. Detection and exploitation of the control switching structure in the solution of dynamic optimization problems. *J. Process Control* **2006**, *16*, 275–290.

19. Schlegel, M. *Adaptive Discretization Methods for the Efficient Solution of Dynamic Optimization Problems*; VDI-Verlag: Dusseldorf, Germany, 2005.

20. Liu, P.; Li, G.; Liu, X.; Zhang, Z. Novel non-uniform adaptive grid refinement control parameterization approach for biochemical processes optimization. *Biochem. Eng. J.* **2016**, *111*, 63–74.

21. Dorfman, R.; Samuelson, P.A.; Solow, R.M. *Linear Programming and Economic Analysis*; McGraw Hill: New York, NY, USA, 1958.

22. McKenzie, L. Turnpike theory. *Econometrica* **1976**, *44*, 841–865.

23. Wilde, R.; Kokotovic, P. A dichotomy in linear control theory. *IEEE Trans. Autom. Control* **1972**, *17*, 382–383.

24. Anderson, B.D.; Kokotovic, P.V. Optimal control problems over large time intervals. *Automatica* **1987**, *23*, 355–363.

25. Rao, A.; Mease, K. Dichotomic basis approach to solving hyper-sensitive optimal control problems. *Automatica* **1999**, *35*, 633–642.

26. Grüne, L.; Müller, M.A. On the relation between strict dissipativity and turnpike properties. *Syst. Control Lett.* **2016**, *90*, 45–53.

27. Faulwasser, T.; Korda, M.; Jones, C.N.; Bonvin, D. On turnpike and dissipativity properties of continuous-time optimal control problems. *Automatica* **2017**, *81*, 297–304.

28. Galán, S.; Feehery, W.F.; Barton, P.I. Parametric sensitivity functions for hybrid discrete/continuous systems. *Appl. Numer. Math.* **1999**, *31*, 17–47.

29. Rawlings, J.; Amrit, R. Optimizing Process Economic Performance Using Model Predictive Control. In *Nonlinear Model Predictive Control*; Magni, L., Raimondo, D., Allgöwer, F., Eds.; Springer: Berlin/Heidelberg, Germany, 2009; Volume 384, pp. 119–138.

30. Carlson, D.; Haurie, A.; Leizarowitz, A. *Infinite Horizon Optimal Control: Deterministic and Stochastic Systems*; Springer: Berlin/Heidelberg, Germany, 1991.

31. Zaslavski, A.J. *Turnpike Properties in the Calculus of Variations and Optimal Control*; Springer: New York, NY, USA, 2006.

32. Zaslavski, A.J. Turnpike Properties of Optimal Control Systems. *Aenorm* **2012**, *20*, 36–40.

33. Zaslavski, A.J. *Turnpike Phenomenon and Infinite Horizon Optimal Control*; Springer: Cham, Switzerland, 2014.

34. Faulwasser, T.; Korda, M.; Jones, C.N.; Bonvin, D. Turnpike and dissipativity properties in dynamic real-time optimization and economic MPC. In Proceedings of the 2014 IEEE 53rd Annual Conference on Decision and Control (CDC), Los Angeles, CA, USA, 15–17 December 2014; pp. 2734–2739.

35. Grüne, L. Economic receding horizon control without terminal constraints. *Automatica* **2013**, *49*, 725–734.

36. Faulwasser, T.; Bonvin, D. On the design of economic NMPC based on approximate turnpike properties. In Proceedings of the 2015 54th IEEE Conference on Decision and Control (CDC), Osaka, Japan, 15–18 December 2015; pp. 4964–4970.

37. Trélat, E.; Zuazua, E. The turnpike property in finite-dimensional nonlinear optimal control. *J. Differ. Equ.* **2015**, *258*, 81–114.

38. Teo, K.L.; Jennings, L.S.; Lee, H.W.J.; Rehbock, V. The control parameterization enhancing transform for constrained optimal control problems. *ANZIAM J.* **1999**, *40*, 314–335.

39. Lee, H.; Teo, K.; Rehbock, V.; Jennings, L. Control parametrization enhancing technique for optimal discrete-valued control problems. *Automatica* **1999**, *35*, 1401–1407.

40. Li, R.; Teo, K.; Wong, K.; Duan, G. Control parameterization enhancing transform for optimal control of switched systems. *Math. Comput. Model.* **2006**, *43*, 1393–1403.

41. Wächter, A.; Biegler, L.T. On the implementation of an interior-point filter line-search algorithm for large-scale nonlinear programming. *Math. Program.* **2006**, *106*, 25–57.

42. Tolsma, J.; Barton, P.I. DAEPACK: An Open Modeling Environment for Legacy Models. *Ind. Eng. Chem. Res.* **2000**, *39*, 1826–1839, doi:10.1021/ie990734o.

43. Tolsma, J.E.; Barton, P.I. Hidden Discontinuities and Parametric Sensitivity Calculations. *SIAM J. Sci. Comput.* **2002**, *23*, 1861–1874, doi:10.1137/S106482750037281X.

44. Tawarmalani, M.; Sahinidis, N.V. A polyhedral branch-and-cut approach to global optimization. *Math. Program.* **2005**, *103*, 225–249.

45. GAMS. GAMS–A User's Guide. Available online: https://www.gams.com/24.8/docs/userguides/GAMSUsersGuide.pdf (accessed on 14 December 2017).

46. Rothfuss, R.; Rudolph, J.; Zeitz, M. Flatness based control of a nonlinear chemical reactor model. *Automatica* **1996**, *32*, 1433–1439.

47. Sahlodin, A.M.; Barton, P.I. Optimal Campaign Continuous Manufacturing. *Ind. Eng. Chem. Res.* **2015**, *54*, 11344–11359.

MDPI AG

St. Alban-Anlage 66

4052 Basel, Switzerland

Tel. +41 61 683 77 34

Fax +41 61 302 89 18

http://www.mdpi.com

Processes Editorial Office

E-mail: processes@mdpi.com

http://www.mdpi.com/journal/processes

www.ingramcontent.com/pod-product-compliance
Lightning Source LLC
Chambersburg PA
CBHW051900210326
41597CB00033B/5965